CHIRAL SEPARATIONS

CHIRAL SEPARATIONS

Edited by

D. Stevenson

The Robens Institute of Industrial and
Environmental Health and Safety
University of Surrey
Guildford, Surrey, United Kingdom

and

I. D. Wilson

ICI Pharmaceuticals
Macclesfield, Cheshire, United Kingdom

PLENUM PRESS • NEW YORK AND LONDON

Library of Congress Cataloging in Publication Data

Chromatographic Society International Symposium on Chiral Separations
(1987: University of Surrey)
  Chiral separations.

  "Proceedings of the Chromatographic Society International Symposium on
Chiral Separations, held September 3–4, 1987, at the University of Surrey,
Guildford, Surrey, United Kingdom"—T.p. verso.
  Bibliography: p.
  Includes index.
  1. Liquid chromatography—Congresses. 2. Enantiomers—Separation—
Congresses. I. Stevenson, D. (Derrick) II. Wilson, Ian D. III. Chromatographic
Society. IV. Title. QD79.C454C485 1987   543'.0894   89-15949

ISBN 978-1-4615-6636-6   ISBN 978-1-4615-6634-2 (eBook)
DOI 10.1007/978-1-4615-6634-2

Proceedings of the Chromatographic Society International Symposium
on Chiral Separations, held September 3-4, 1987, at the
University of Surrey, Guildford, Surrey, United Kingdom

© 1988 Plenum Press, New York
A Division of Plenum Publishing Corporation
233 Spring Street, New York, N.Y. 10013
Softcover reprint of the hardcover 1st edition 1988

PREFACE

This volume represents the proceedings of a two-day international meeting on chiral chromatography held at the University of Surrey between 3-4 September 1987. The meeting was jointly organized by the Chromatographic Society and the Robens Institute of the University of Surrey in response to the burgeoning interest in this rapid maturing field of chromatography. Nowhere is this interest more evident than in the agrochemical and pharmaceutical industries where the implications of different pharmacological and toxicological activity for the individual enantiomers present in a racemic drug or insecticide is an increasing area of concern. Developments in the area of chiral separations are at last beginning to provide scientists with the necessary tools to study how animals and man handle racemates and relate their observations to the observed biological effects of these substances. The development of robust and simple methods for the separation of enantiomers will therefore have a profound impact on safety evaluation and drug design.

The meeting proved to be very successful, with over 160 delegates from thirteen countries in Europe and America present to learn from the experiences of experts in the field of chiral chromatography and to hear about the latest developments. Hopefully, in future symposia on chiral separations at the University of Surrey, some of the delegates who were at this first meeting mainly to learn will return to report on their use of chiral chromatography in one of the most difficult areas facing modern analysts — the reliable and accurate quantification of enantiomers at trace concentrations in biological samples.

D. Stevenson and I.D. Wilson

CONTENTS

The Chromatographic Society   ix

The Robens Institute   xi

The Biological Importance of Chirality and Methods Available to
Determine Enantiomers
D. Stevenson and G. Williams   1

Drug Analysis Using High-Performance Liquid Chromatographic (HPLC)
Chiral Stationary Phases: Where to Begin and Which to Use
Irving W. Wainer, Rose M. Stiffin and Ya-Qin Chu   11

Systematic Studies of Chiral Recognition Mechanisms
William H. Pirkle, Thomas C. Pochapsky, John A. Burke III
and Kris C. Deming   23

Separation of Enantiomers of Oxyphenonium Bromide by High-Performance
Liquid Chromatography
Karla G. Feitsma, Ben F.H. Drenth and Rokus A. de Zeeuw   37

The Use of Pirkle High-Performance Liquid Chromatography Phases in the
Resolution of Enantiomers of Polycyclic Aromatic Hydrocarbon
Metabolites
Michael Hall and Philip L. Grover   43

The Role of Solvents and Steric Factors in the Resolution of β-Blocker
Drugs on Chiral Urea Phases
Nagaraja K.R. Rao, Robert C. Towill and Bindu Todd   55

Complexation of Dansyl Amino Acid Enantiomers by β-Cyclodextrin
Studied by High-Performance Liquid Chromatography and
Fluorescence Measurements
David A. Briggs, Roger B. Homer and Russell Godfrey   61

Analytical and Preparative Chiral Resolution of Some Aminoalcohols
by Ion-Pair High-Performance Liquid Chromatography
R.M. Gaskell and B. Crooks   65

Simultaneous Enantioselective Determination of Underivatized
β-Blocking Agents and their Metabolites in Biological Samples
by Chiral Ion-Pairing High-Performance Liquid Chromatography
T. Leeman and P. Dayer   71

Chiral Gas and High-Performance Liquid Chromatographic Analysis of
        Enantiomers of Fungicides and Plant Growth Regulators: Application
        in Fungal, Plant and Soil Metabolism Studies
    T. Clark, A.H.B. Deas and K. Vogeler   79

Recent Developments in Enantiomer Separation by Complexation Gas
        Chromatography
    Volker Schurig and Rainer Link   91

Use of Various Commercially Available Chiral Stationary Phases in
        Supercritical Fluid Chromatography
    P. Macaudiere, M. Caude, R. Rosset and T. Tambute   115

Strategies for Optimising Chiral Separations in Drug Analysis
    Anthony F. Fell and Terence A.G. Noctor   121

Further Use of Computer Aided Chemistry to Predict Chiral Separations
        in Liquid Chromatography: Selecting the Most Appropriate
        Derivative
    Ulf Norinder and E. Goran Sundholm   127

An Optical Rotation Detector for High-Performance Liquid Chromatography
    D.M. Goodall and D.K. Lloyd   131

Prospects for Chiral Thin-Layer Chromatography
    I.D. Wilson and R.J. Ruane   135

Appendix 1 — Chiral Chromatography Literature 1987-1988   145

Appendix 2 — Some Manufacturers and Suppliers of Chiral Columns   179

Appendix 3 — Abstracts   181

Author Index   197

Compound Index   199

Subject Index   203

THE CHROMATOGRAPHIC SOCIETY

The Chromatographic Society is the only international organization devoted to the promotion of, and the exchange of information on, all aspects of chromatography and related techniques.

With the introduction of gas chromatography in 1952, the Hydrocarbon Chemistry Panel of the Hydrocarbon Research Group of the Institute of Petroleum, recognizing the potential of this new technique, set up a Committee under Dr. S. F. Birch to organize a Symposium on 'Vapor Phase Chromatography' which was held in London in June 1956. Almost 400 delegates attended this meeting and success exceeded all expectation. It was immediately apparent that there was a need for an organized forum to afford discussion of development and application of the method and, by the end of the year, the Gas Chromatography Discussion Group had been formed under the Chairmanship of Dr. A. T. James with D. H. Desty as Secretary. Membership of this Group was originally by invitation only but, in deference to popular demand, the Group was opened to all willing to pay the modest subscription of one guinea and in 1957 A.J.P. Martin, Nobel Laureate, was elected inaugural Chairman of the newly-expanded Discussion Group.

In 1958 a second Symposium was organized, this time in conjunction with the Dutch Chemical Society, and since that memorable meeting in Amsterdam the Group, now Society, has maintained close contact with kindred bodies in other countries, particularly France (Groupement pour l'Avancement des Methodes Spectroscopiques et Physico-chimiques d'Analyse) and Germany (Arbeitskreis Chromatographie der Gesellschaft Deutscher Chemiker) as well as interested parties in Eire, Italy, The Netherlands, Scandinavia, Spain and Switzerland. As a result Chromatography Symposia, in association with Instrument Exhibitions, have been held biennially in Amsterdam, Edinburgh, Hamburg, Brighton, Rome, Copenhagen, Dublin, Montreux, Barcelona, Birmingham, Baden-Baden, Cannes, London, Nurnburg, Paris and Vienna.

In 1958 'Gas Chromatography Abstracts' was introduced in journal format under the Editorship of C.E.H. Knapman; first published by Butterworths, then by the Institute of Petroleum, it now appears as 'Gas and Liquid Chromatography Abstracts' produced by Elsevier Applied Science Publishers and is of international status — abstracts, covering all aspects of chromatography, are collected by Members from over 200 sources and collated by the Editor Mr. E. R. Adlard assisted by Dr. P. S. Sewell.

Links with the Institute of Petroleum were severed at the end of 1972 and the Group established a Secretariat at Trent Polytechnic in Nottingham, Professor Ralph Stock playing a prominent part in the establishment of the Group as an independent body. At the same time, in recognition of expanding horizons, the name of the organization was changed to the Chromatography Discussion Group.

In 1978, the 'Father' of Partition Chromatography, Professor A. J. P. Martin was both honoured and commemorated by the institution of the Martin Award which is designed as testimony of distinguished contribution to the advancement of chromatography. Recipients of the award include:-

E.R. Adlard	Prof. E. Bayer
Professor U.A.Th. Brinkman	Dr L.S. Ettre
Professor J.C. Giddings	Professor G. Guiochon
Professor J.F.K. Huber	Dr C.E.R. Jones
C.E.H. Knapman	Professor J.H. Knox
Dr E. Kovats	Professor A. Liberti
Dr C.S.G. Phillips	Dr G. Schomburg
Dr R.P.W. Scott	Professor R. Stock
and Dr. G.A.P. Tuey	

The Group celebrated its Silver Jubilee in 1982 with the 14th International Symposium held, appropriately, in London. To commemorate that event the Jubilee Medal was struck as means of recognising significant contributions by younger workers in the field. Recipients of the Jubilee Medal include: Dr H. Colin, Dr K. Grob Jr., Dr J. Hermannson, Dr P.G. Simmonds and Dr R. Tijssen.

In 1984 the name was once again changed, this time to The Chromagraphic Society, which title was believed to be more in keeping with the role of a learned society having an international membership of some 1000 scientists drawn from more than 40 countries. At that time, the Executive Committee instituted Conference and Travel Bursaries in order to assist Members wishing to contribute to, or attend major meetings throughout the world.

The Society is run by an Executive Committee elected by its Members, in addition to the international symposia, seven or eight one-day meetings covering a wide range of subjects are organized annually. One of these meetings, the Spring Symposium, is coupled with the Society's Annual General Meeting when, in addition to electing the Society's Executive Committee, Members have the opportunity to express their views on the Society's activities and offer suggestions for future policy.

Regular training courses in all aspects of chromatography are run in conjunction with the Robens Institute of the University of Surrey and it is hoped that this particular service will eventually include advanced and highly specialised instruction.

Reports of the Society's activities, in addition to other items of interest to its members (including detailed summaries of all papers presented at its meetings), are given in the Chromatographic Society Bulletin which is produced quarterly under the editorship of I.W. Davies.

At the time of writing three grades of membership are offered:

Membership with Abstracts	£20.00 per year
Membership	£11.00 per year
Student Membership (includes Abstracts)	£6.00 per year

Members receive the Bulletin free of charge, benefit from concessionary Registration Fees for all Meetings and Training Courses and are, of course, eligible to apply for Travel and/or Conference Bursaries.

For further information please write to:

Mrs. J. Challis
Executive Secretary
THE CHROMATOGRAPHIC SOCIETY
Trent Polytechnic
Burton Street
Nottingham NG1 4BU
United Kingdom.

THE ROBENS INSTITUTE OF INDUSTRIAL AND

ENVIRONMENTAL HEALTH AND SAFETY

The Robens Institute of Industrial and Environmental Health and Safety was established by the University of Surrey in 1978. It is housed on the University campus at Guildford. The Institute carries out both academic and contract research in health and safety related areas, including analytical chemistry, toxicology, occupational health and hygiene, ergonomics and environmental health. In addition, The Robens runs a major programme of post-experience courses and symposia. In 1984 the Chromatographic Society set up a series of training courses (in GLC and HPLC) now run annually at The Robens.

CHIRAL SEPARATIONS

THE BIOLOGICAL IMPORTANCE OF CHIRALITY AND METHODS

AVAILABLE TO DETERMINE ENANTIOMERS

D. Stevenson and G.A. Williams

The Robens Institute of Industrial and Environmental Health and Safety
University of Surrey, Guildford, Surrey GU2 5XH, UK

SUMMARY

Many compounds are marketed as racemic mixtures even though it is now known that enantiomers can have different biological activities. This can be due to differences in, for example, protein binding, transport, metabolism and clearance. Methods available to determine enantiomers include chromatography on chiral stationary phases, the use of chiral additives in the mobile phase, derivatization to yield diastereoisomers, chiral detectors, NMR and enantiomer specific immunoassays. The advantages of the various methods are discussed.

THE BIOLOGICAL IMPORTANCE OF CHIRALITY

Compounds that exist in two forms that are nonsuperimposable mirror images show optical activity. That is, they rotate the plane of plane-polarised light in opposite directions. This property is shown by an asymmetric carbon atom (ie. one with four different substituents) but also by other atoms such as sulphur, phosphorus, some metal atoms and can also occur when rotation around an atomic bond is hindered by bulky functional groups. Compounds differing only in their ability to rotate plane-polarised light in opposite directions are known as enantiomers.

Many compounds, such as drugs, agrochemicals and food additives have been marketed as racemic mixtures (that is approximately equal proportions of enantiomers) even though it has long been known that different enantiomers may show very different biological activity. Differences in biological activity of drugs, agrochemicals, etc., may arise due to differences in:

- protein binding and transport [1]
- mechanism of action [2]
- rates of metabolism
- changes in activity due to metabolism
- clearance rates [3]
- persistence in the environment

It must be remembered that the human body is a highly stereospecific environment (ie. we have D-sugars and L-amino acids) and that in nature, asymmetry at the molecular level is the norm, not the exception[4]. Regulatory authorities have become aware that the efficacy and toxicology of enantiomers may differ from each other and from the racemic mixture, and are increasingly asking for data on stereochemistry.

1

It has also been discovered that metabolism may produce chiral metabolites from non-chiral molecules[5]. Some examples are shown below

Ketone reduction

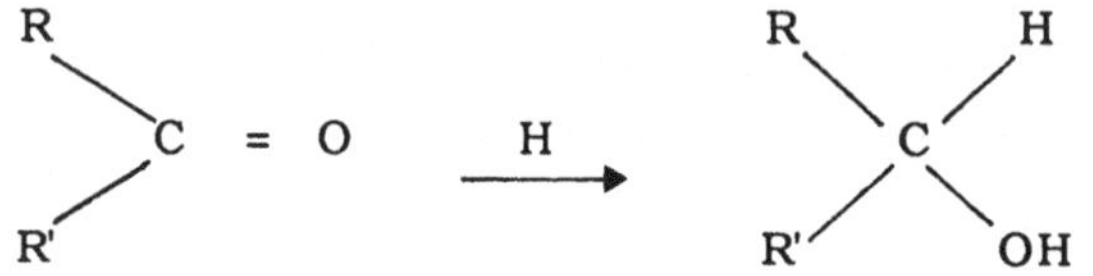

```
  R                          R    H
   \                          \  /
    C = O      --H-->          C
   /                          / \
  R'                         R'  OH
```

Reduction of a double bond

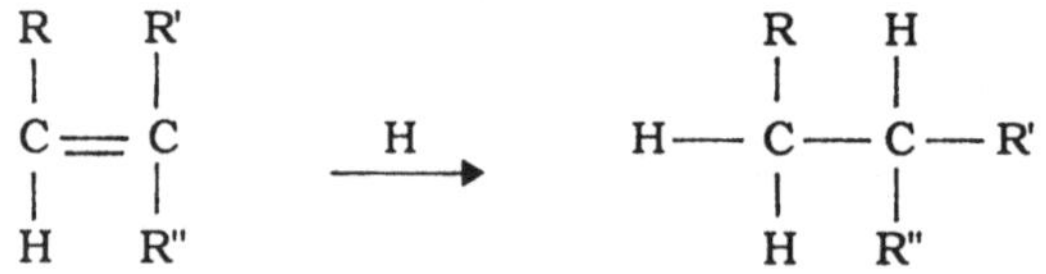

```
  R    R'                    R   H
  |    |                     |   |
  C == C       --H-->    H — C — C — R'
  |    |                     |   |
  H    R"                    H   R"
```

Hydroxylation

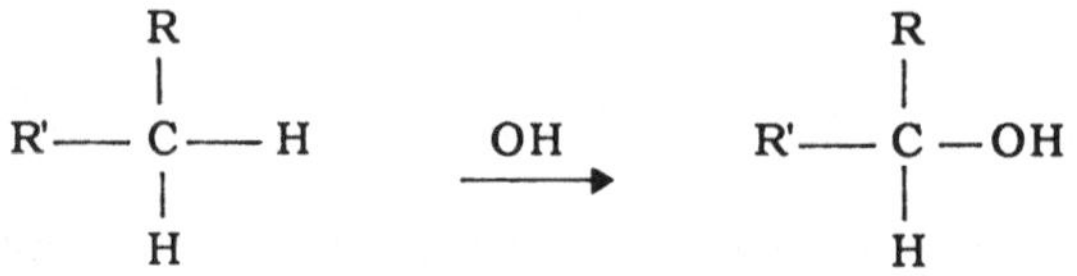

```
       R                          R
       |                          |
  R' — C — H     --OH-->    R' — C — OH
       |                          |
       H                          H
```

Oxidation

```
       R                          R
       |                          |
  R' — C — X      --O-->     R' — C — OX
       |                          |
       X                          X
```

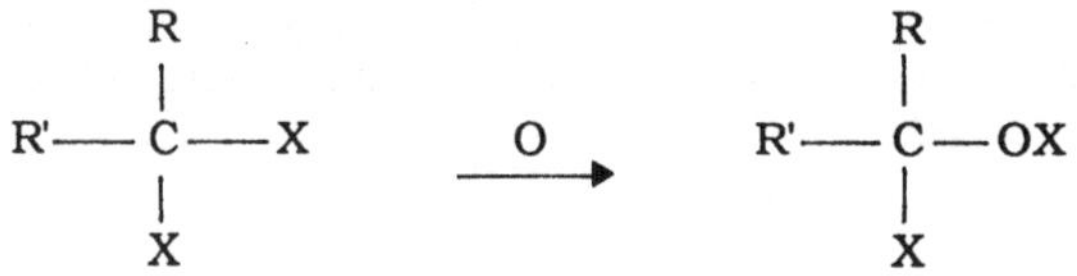

Oxidation of tertiary amine

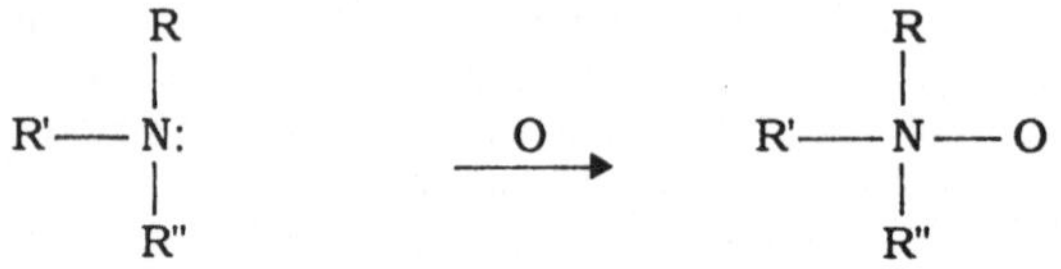

```
       R                          R
       |                          |
  R' — N:         --O-->     R' — N — O
       |                          |
       R"                         R"
```

The importance of chirality when investigating the mechanism of toxicity can be demonstrated with the general example below. The toxic effect of a hydrocarbon was thought to be caused by a metabolite.

```
       H                         OH
       |                         |
  R —— C —— R'      -->    R —— C —→ R'
       |                         |
       H                         H
```

The metabolite was isolated, characterized and synthesized for toxicity testing for comparison with the pattern seen for the parent compound. The activity of the chemically synthesized (racemic) metabolite was different to the *in-vivo* formed metabolite when one enantiomer predominated.

It is clear, therefore, that restricting the study of chiral aspects of metabolism to comparing the fate of enantiomers in biological fluid, may be misleading. It has also been discovered that enzymes capable of converting one enantiomer to its antipode exist, as has been shown with ibuprofen.

Analytical methods capable of discriminating between enantiomers are thus required for a variety of reasons, including:

- checking the purity of products where the aim is to market only one enantiomer
- monitoring asymmetric synthesis
- determination of enantiomeric drugs in body fluids for pharmacokinetic studies.
- in-vivo metabolism studies
- in-vitro metabolism studies
- checking if the interconversion pathway is active for a particular compound
- testing the fate of agrochemicals in the environment.

METHODS AVAILABLE FOR CHIRAL DISCRIMINATION

The separation of enantiomers is a very exacting analytical task but not a new one. In 1848 Louis Pasteur carried out mechanical (ie. hand and eye) separation of enantiomers and then in 1858 used bacteria to discriminate between enantiomers by selectively destroying one. At about that time Pasteur also used a chiral reagent to form diastereoisomeric salts which have different physical properties, thus allowing separation by fractional crystallization. In the 1930s Henderson and Rule used the selective adsorbtion of camphor derivatives onto D-lactose as a means of chiral discrimination. In the 1960s Gil-Av and others used gas liquid chromatography (GLC) with chiral stationary phases and also derivatization with chiral reagents followed by achiral GLC or thin-layer chromatography (TLC). In the 1960s high performance liquid chromatography (HPLC) was used after derivatization with chiral reagents, then in the 1980s HPLC with chiral stationary phases such as those developed by Pirkle became available. The modern analyst therefore has several different approaches that can be tried for the separation and/or determination of enantiomers

- Derivatization using chiral reagents then HPLC, GLC, TLC
- Chiral stationary phases for HPLC, GLC, TLC
- Chiral mobile phases for HPLC and TLC
- Chiral detectors for HPLC
- Nuclear magnetic resonance (NMR) using chiral shift reagents
- Enantiomer specific immunoassays.

DERIVATIZATION USING CHIRAL REAGENTS

Until the advent of chiral stationary phases, the most common means of achieving the separation of enantiomers was by reaction with a reagent containing another asymmetric centre to form a pair of diastereoisomers. As diastereoisomers differ in their physical properties, these can be separated by conventional (non-chiral) stationary phases. In order to use this approach successfully, the reagent must be readily available in a chemically and optically pure form, must react at the same rate with both enantiomers and racemisation, decomposition and side reactions must not occur during the derivatization. Clearly, the analyte molecules must contain a functional group suitable for derivatization and for practical reasons it is preferable that the reaction is quick. Examples of reagents used include:

- for aminoacids	N-trifluoroacetyl prolylchloride
	α - chloroisovalenylchloride
	menthylchloroformate
- for amines	N-trifluoroacetyl prolylchloride
	α - phenylbutyric anhydride
- for alcohols	2 - phenylpropionylchloride
	1 - phenylethylisocyanate menthyl chloroformate

- for aliphatic	desoxyephedrine l-menthol
and alicyclic acids	
- for ketones	2,2,2-trifluoro-1-pentylethylhydrazine

An excellent list of chiral derivatization reagents is given by Souter 1985[6].

When using derivatization for HPLC it is also possible to introduce a chromophore to enhance detection; for example, using (+)-1-(9-fluoroenyl) ethylchloroformate[7]. The advent of chiral stationary phases, particularly for HPLC means that this approach is usually favored ahead of diastereoisomer formation. However, chiral chromatography is often unsuccessful and so diastereoisomer formation will continue to receive attention[8] as a quick and robust solution. Enantiomer resolution using this approach is usually enhanced when bulky groups are attached to the chiral centre and when the chiral centres of both the reagent and the analyte are in close proximity in the resulting diastereoisomer.

CHIRAL STATIONARY PHASES FOR HPLC, GLC AND TLC

HPLC

Presently, the use of HPLC with chiral stationary phases is the most common 'first approach' for enantiomer separation. A large number of different phases are commercially available; all are fairly expensive and, in some case, of questionable stability and reproducibility from column-to-column. In many ways this mirrors the early development of bonded phases for HPLC and hopefully the situation will improve. Most HPLC chiral stationary phases have quite specific structural requirements for successful enantiomer separation and therefore apply only to a limited range of compounds. Several authors have classified columns into groups and have attempted to describe the mechanism of separation and thus the structural requirements on the analyte [9-12].

i)   *Donor-Acceptor (Pirkle) Type Phases*

The first commercially available chiral stationary phase for HPLC was R-N-(3,5 - dinitrobenzoyl) phenylglycine bound ionically to aminopropyl silica[13]. Several other similar phases have followed. The structure of this phase is shown in Fig.1. It has four possible points of interaction with analyte molecules, three are necessary for chiral separation to occur. These interactions can include hydrogen-bonding, dipole-stacking, charge-transfer and steric effects. The π-acceptor dinitrobenzoyl group can attract π-donor groups such as aromatic rings or alkyl groups. Many examples of separations on these columns have been reported often requiring achiral derivatization to reduce analyte polarity (a disadvantage of these phases) or to introduce functional groups suitable to interact with the phase.[14]. With the Pirkle type columns, both ionic and covalently bound phases are available. These often give similar results though occasionally selectivity is different. The ionic phases can only be used with organic eluents (even then not more than ~20% propan-2-ol in hexane). Sometimes the columns can be regenerated using commercially available solutions of the stationary phases.

ii)   *Chiral Cavity Phases*

Chiral cavity type phases such as cyclodextrins bound to silica through a spacer are now also commercially available and in widespread use[15,16]. The cyclodextrin structure can be viewed as a cone (Fig.2) open at both ends with the interior of the cone relatively hydrophobic. If a chiral molecule fits exactly into the cone and has functional groups that interact with the secondary hydroxyl groups at the wider opening of the cone, then chiral separation may occur. Different cyclodextrin and derivatized cyclodextrins with different sized chiral cavities are available. Small molecules totally included in the cone will not separate. The cyclodextrin columns are used in the reversed-phase mode using typical water/methanol or acetonitrile type mobile phases, the organic modifier competes with solute molecules for inclusion in the hydrophobic cavity.

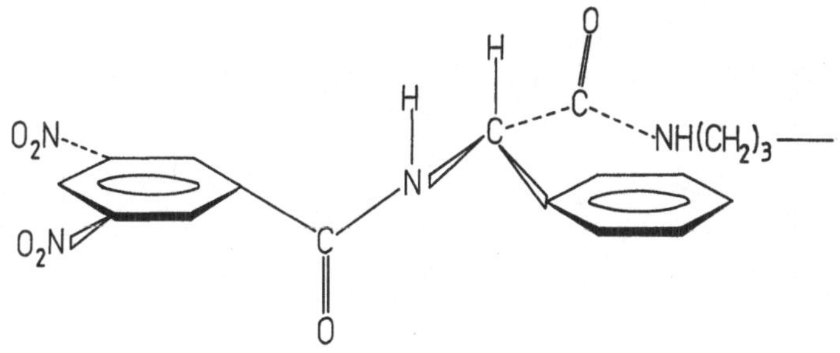

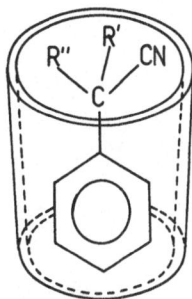

Fig. 1. The structure of 3, 5-dinitrobenzoyl phenyl glycine, used in Pirkle type chiral stationary phases for HPLC.

Fig. 2. Inclusion of analyte molecule in a cyclodextrin 'chiral cavity' to give chiral separation.

Mobile phase additives such as buffer salts can help improve column efficiency and resolution. The cyclodextrin columns are reactively cheap. Cyclodextrin bonded phases are also very useful for separating cis-trans isomers.[17].

iii)   *Helical Polymer Phases*

A third class of chiral stationary phase are helical polymers such as cellulose esters (Fig.3) and poly-(triphenylmethlmetacrylates) whose chirality arises from helicity[18-20]. It is thought that the mechanism of separation on these columns involves a combination of attractive interactions and inclusion of the analyte in a chiral cavity. They are mostly used in the normal phase mode and can be used as stationary phases themselves or as lower molecular weight fractions adsorbed onto silica gel. These columns are relatively expensive and are also subject to pressure limitations.

iv)   *Ligand Exchange Columns*

Ligand exchange columns such as those developed by Davankov have been used for chiral separations for many years[21]. Stationary phases such as proline or hydroxyproline bound to silica via a spacer are commercially available. The phases are treated with copper salts. In order to achieve chiral separation the analyte molecules must form a reversible complex with the copper ion.[22] (Fig.4). The analyte molecules must have two polar functional groups on (or adjacent to) the asymmetric carbon atom. Thus α-amino acids and amino alcohols are ideal for this type of column, but they are not applicable for other compound types. They are used with aqueous mobile phases.

v)   *Protein Phases*

The fifth type of chiral stationary phases are proteins bound to silica. Two are commercially available; bovine serum albumin (BSA) [23,24] and alpha acid glycoprotein (AGP)[25,26]. These are used with aqueous buffers (and small amounts of organic modifers)

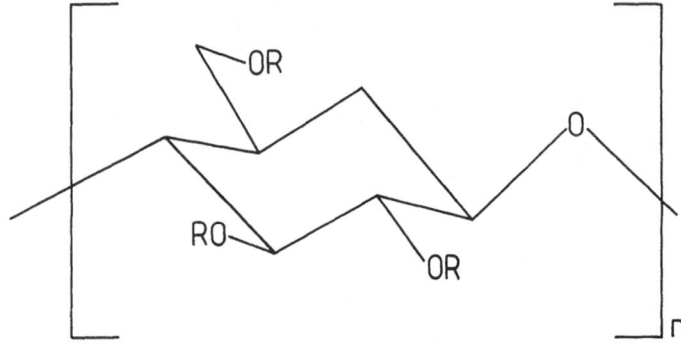

Fig. 3. Basic unit of cellulose type phase.

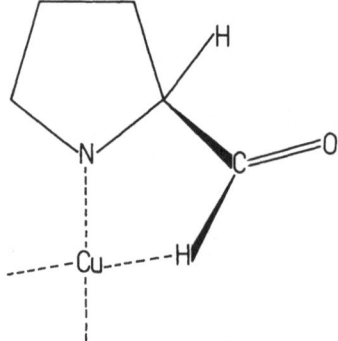

Fig. 4. Ligand exchange type phases requiring complexing with Cu ions.

as the proteins are not stable with organic solvents or extremes of pH. These columns can give very good resolution of enantiomers through chromatographic efficiency is often poor. Separations are thought to be based on combinations of hydrophobic and polar interactions, although ionic interactions may also occur, structural requirements for chiral separation are therefore difficult to predict. Generally speaking, the protein columns are of low capacity and cannot be used for preparative work.

A summary of some of the advantages and disadvantages of HPLC chiral stationary phases is shown in the table[27].

GLC

The direct resolution of enantiomers on chiral stationary phases is also possible,[28,29] though there are very few such phases commercially available. The most commonly used phase is the capillary column Chirasil-Val® (both D and L are available).The structure of this phase which is based on L-valine-tert butylamide is shown in Fig.5. The phase has been used to separate many drugs, hydroxy acids, amino acids and amino alcohols. Enantiomer separation by GLC is much less common than HPLC The technique is limited to relatively volatile analytes or their derivatives. The Chirasil-Val phase itself cannot be used above temperatures of about 230°C as racemization of the stationary phase can occur.

TLC

Chiral stationary phases for thin-layer chromatography are also known. The Chiraplate® made by treating octadecyl-modified silica with copper acetate and 4 hydroxy-1-(2-hydroxydodecyl) proline as a chiral selector is commercially available. These plates have been used to separate underivatized amino acids[30]. Separation is based on ligand exchange and only molecules possessing two functional groups on (or possibly adjacent to) the asymmetric carbon atom capable of complexing with the copper ion will separate. In

6

Table 1.   Some Advantages and Disadvantages of Chiral Stationary Phases for HPLC

Ligand Exchange Columns
Use aqueous mobile phases
Require two polar functional groups on or near chiral centre

Cyclodextrins
Use aqueous mobile phases
Fairly cheap

Pirkle Type Phases
Often require derivatization
Wide range of applications with specific interactions
Need to use non-polar mobile phases for ionic columns

Helical Polymers
Pressure limitations
Solvent limitations

Protein Columns
Use hydrophobic and polar interactions, multiple binding sites
Use aqueous mobile phase pH 5-9
Tolerate only small amounts of organic modifier, column efficiency low

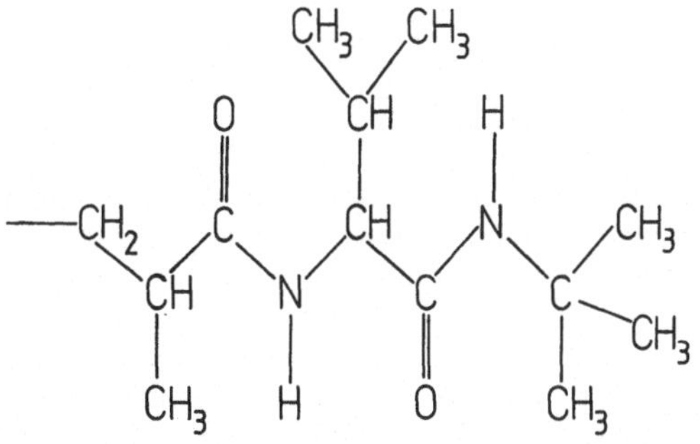

Fig. 5. The structure of Chirasil-Val.

the authors' laboratory, separations were achieved in an unsaturated chamber with a development time of about 15 minutes for a distance of 10 cm. The mobile phases were mixtures of methanol/water/acetonitrile. The plates were easy to use and reproducible. Other stationary phases analogous to the HPLC phases such as Pirkle type phases and cyclodextrins have been reported[31] and may soon become commercially available. Chiral TLC could prove useful in metabolism studies and also as a guide to selection of highly expensive HPLC phases.

CHIRAL MOBILE PHASE ADDITIVES

The separation of enantiomers on chiral stationary phases involves the formation of reversible diastereomeric complexes. The same effect can sometimes be achieved by adding chiral reagents to the mobile phase and using non-chiral stationary phases.[32,33]. Optically active ion-pair reagents have been used, particularly for basic drugs using pairing ions such as camphor sulphonic acid and Z-glycyl proline. Bovine serum albumin and cyclodextrin [34] (both more commonly used as a bound stationary phase) have also been used successfully in the mobile phase. In some cases it is quite likely that separation occurs via '*in-situ*' coating of the column to form a temporary chiral stationary phase. Chiral mobile phase additives have the advantage that less expensive column packings can be used and that there is an even wider choice than of stationary phases. They can also be used for preparative isolation, for example, when using chiral ion-pairing reagents. Mobile phase additives are limited to compounds that will not affect the detection system, and the addition of large amounts of some additives can prove costly.

CHIRAL DETECTORS FOR HPLC

The traditional means of distinguishing between enantiomers spectroscopically has involved using plane polarised light and measuring optical rotation. This is a simple method, ideal when dealing with bulk quantities of enantiomers of known specific rotation. A racemic mixture will give zero specific rotation. The technique at present is not sensitive enough for trace analysis, though sensitivity does vary from compound to compound Recently polarimetric detectors specifically designed for HPLC have become available[35,36] often used in series with a UV or refractive index detector to give both quantification and the ratio of enantiomers. A limit of detection as low as 10 ug is now possible. Such detectors allow chiral discrimination without chromatographic separation, excess of one enantiomer giving peaks in the opposite direction to the other.

NMR

Nuclear Magnetic Resonance (NMR) can often distinguish between diastereoisomers and can therefore be used directly after derivatization with a chiral reagent. More convenient is the use of chiral shift reagents which can also give a different spectrum for two enantiomers. NMR does lack sensitivity but is able to give much more structural information than chromatographic techniques (unless these are used in combination with mass-spectrometry).

ENANTIOMERIC SPECIFIC IMMUNOASSAYS

Immunoassays are often used as an alternative to chromatrographic methods, particularly for the trace determinations of drugs and endogenous compounds in biological fluids. They can have advantages of sensitivity and simplicity. It has also been demonstrated that immunoassays can distinguish between enantiomers.[37]. Immunoassays used for the determination of drugs in biological fluids sometimes suffer from cross-reactivity of closely related compounds such as metabolites. Immunoassays are thus quite an attractive proposition for the determination of drugs where metabolism is absent and maybe also for bulk drug formulations where a small amount of one enantiomer 'impurity' is present.

CONCLUDING REMARKS

The significance now attached to the different biological activities of enantiomers has presented a very exacting task to the modern analytical chemist. Enantiomer separation has now been achieved for a wide range of compounds, particularly by HPLC. However, because of growing regulatory concern it is likely that more drugs, etc., will be marketed as the single enantiomer. Methods to separate a small amount of the enantiomer in a large excess of the other, will thus be required for purity checking. The very many chiral separations now shown for pure compounds are not necessarily adequate for this type of quantification as chromatographic efficiency is often not good for chiral stationary phases. There are other practical limitations with many phases and each one is applicable for only a small range of compounds. The development of new chiral stationary phases is thus likely to continue. Hopefully, molecular graphics and/or a better understanding of separation mechanisms will allow us to predict with greater certainty which phase to use for a particular separation. It is also likely that non-chromatographic methods such as immunoassays and NMR and the use of chiral detectors will also receive more attention.

As more measurements of enantiomers are made, a general understanding of the implications of chirality will be achieved. This could lead to better therapeutic drug monitoring, a better understanding of inter-individual and species' differences in xenobiotic metabolism. So far, most attention has been focused on drugs. It is also likely that other compounds to which the population at large is exposed will also come under scrutiny.

REFERENCES

1. Simonyi, M., Fitos, I., and Vizy, J., *TIPS*, 7 (1986) 112.
2. Lehmann, F.P.A., *TIPS*, 7 (1986) 281.
3. Walle, T., and Walle, K.W. *TIPS*, 7 (1986) 155.
4. Mason, S., *TIPS*, 7 (1986) 20.
5. Testa, B., *TIPS*, 7 (1986) 60.
6. Souter, R.W. (ed) Chromatographic Separation of Stereoisomers (1985), CRC Press, Boca Raton.
7. Einarsson , Josefsson, B., Moller, P., and Sanchez, D., *Anal.Chem.*, 59 (1987) 1191.
8. Gal. 3., *LC - GC*, 5 (1987) 106.
9. Wainer, I.W., *Trends in Anal.Chem.*, 6 (1987) 125.
10. Amstrong , D.W., *Anal.Chem.*, 59 (1987) 84A.
11. Pirkle, W.H. and Pochapsky, T.C., *Adv. Chromatogr.*, 27 (1987) 73.
12. Wainer, I.W., and Doyle, T.D., *HPLC*, 2 (1984) 88.
13. Pirkle, W.H., Finn, J.M., Schreiner, J.L., and Hamper, B.C., *J.Am.ChemSoc.*, 103 (1981) 3964.
14. Wainer, I.W., Dogle, T.D., and Breder, C.D., *J.Liq.Chromatogr.*, 7 (1984) 731.
15. Hinze, W.L., Riehl, T.E., Armstrong, D.W., Demond, W., Alak, A., and Ward, T., *Anal.Chem.*, 57 (1985) 237.
16. Amstrong, D.W., Ward, T.J., Armstrong, R.D., and Beesley, T.E., *.Sci.*, 232 (1986) 1132.
17. Amstrong, D.W., and DeMond, W., *J.Chromatogr.Sci.*, 22 (1984) 411.
18. Ichida, A., Shabata, T., Okamoto, I., Yuki, Y., Namikashin, N., and Toga, Y., *Chromatographia*, 19 (1984) 280.
19. Lindner, K.R., and Mannschreck, A., *J.Chromatogr.*, 193 (1980) 308.
20. Dappen, R., Arm, H., and Meyer, V.R., *J.Chromatogr.*, 373 (1986) 1.
21. Davankov, V.A., *Adv.Chromatogr.*, 18 (1980) 139.
22. Gubitz, G., *J.Liq.Chromatogr.*, 9 (1980) 519.
23. Allenmark, S., Bomgreen, B., and Boren, H., *J.Chromatogr.*, 264 (1983) 63.
24. Allenmark, S., Bomgreen, B., and Boren, H., *J.Chromatogr.*, 316 (1984) 617.
25. Hermansson, J., *J.Chromatogr.*, 269 (1983) 71.
26. Hermansson, J., and Ericksson, M., *J.Liq.Chromatogr.*, 9 (1986) 62.
27. Dappen, R., Arm, H., and Meyer, V.R., *J.Chromatogr.*, 373 (1986) 1.
28. Konig, W.A., (ed) The Practice of Enantiomer Separation by Capillary Gas Chromatography, (1987) Huthig, New York.
29. Liu, R.H., and Ku, W.W., *J.Chromatogr.*, 271 (1983) 309.
30. Brinkman, U.A.Th., and Kamminga, D., *J.Chromatogr.*, 330 (1985) 375.
31. Wainer, I.W., Brunner, C.A., and Doyle, T.D., *J.Chromatogr.*, 264 (1983) 154.
32. Pettersson, C., and Schill, G., *J.Liq.Chromatogr.*, 9 (1986) 269.
33. Pettersson, C., *Eur.Chrom.News*, 2 (1988) 16.
34. Debowski, J., Sybilska, D., and Jurczak, J., *J.Chromatogr.*, 237 (1982) 303.
35. Drake, A.F., Gould, J.M., and Mason, S.F., *J.Chromatogr*, 202 (1980) 239.
36. Scott, B.S., and Dunn, D.L., *J.Chromatogr.*, 202 (1980) 239.
37. Kawashima, K., Levy, A., and Spector, S., *J.Pharmacol.Exp.Ther.*, 196 (1976) 517.

DRUG ANALYSIS USING HIGH-PERFORMANCE LIQUID CHROMATOGRAPHIC (HPLC)

CHIRAL STATIONARY PHASES: WHERE TO BEGIN AND WHICH TO USE

Irving W. Wainer*, Rose M. Stiffin and Ya-Qin Chu**

Pharmaceutical Division
St. Jude Children's Research Hospital
Memphis, TN 38101, USA

SUMMARY

The most common approach to the resolution of enantiomeric compounds involves the conversion of the enantiomers into diastereomers. Unlike enantiomers, diastereomers have different physical properties and can often be easily separated. Using this approach, enantiomers can be resolved through the formation of diastereomeric salts, by the synthesis of diastereomeric derivatives or through the formation of transient diastereomeric complexes. The latter approach is the basic mechanism of HPLC chiral stationary phases (HPLC-CSPs) where transient diastereomeric complexes are formed between enantiomeric solute molecules and an optically pure molecule or polymer on the stationary phase.

In 1981, the first commercially available HPLC-CSP was introduced and the chromatographer was presented with the situation of adapting the analysis to meet the requirements of the column. At the present time the situation is reversed. With over 35 commercially available HPLC-CSPs the chromatographer has the ability to select an HPLC-CSP to fit the analysis. The problem is how to choose the right one. A possible solution to this problem is a proposed method for the classification of HPLC-CSPs. In this approach, the phases are divided into 5 classes according to the chiral recognition process which operates on the CSP. This approach can be used to provide a guide to enable the separation of enantiomeric amino acids, sulfoxides and carboxylic acids.

INTRODUCTION

The ability to rapidly and accurately determine the stereochemical composition, especially the enantiomeric composition, of drug substances is becoming increasingly important in their development, regulation and clinical application. This interest is due to the fact that enantiomeric molecules often differ in potency, toxicity, pharmacological action, metabolism, plasma disposition and urine excretion kinetics.

---

* To whom correspondence should be addressed
** On leave from The National Institute for the Control of Pharmaceutical and Biological Products, Ministry of Health, Beijing, China and supported in part by The Rockefeller Foundation (RF 86068 ~A54).

The pharmacokinetic and pharmacodynamic parameters of the enantiomers of a chiral drug substance are usually determined by studying each enantiomer separately. While an important step in the determination of these parameters, it often gives a false picture of the situation after the adminstration of a racmic mixture of the two enantiomers. This has been recently demonstrated by Giacomini, et al.[1] in their study of the *in vivo* interaction of the enantiomers of disopyramide in humans. Disopyramide is an antiarrhythmic agent which is marketed as a racemeate. When the enantiomers are administered separately, the plasma clearance, renal clearance and volume of distribution are the same for both isomers. However, when the racemic mixture is adminstered, there is a marked difference between the isomers. In this case, *d*-disopyramide has a lower plasma and renal clearance, a longer half-life and a smaller apparent volume of distribution than *l*-disopyramide.

The current pharmacological understanding is that a racemic drug must be treated as a mixture of two separate drugs which requires a complete investigation of the pharmacological properties of both isomers. This realization has resulted in an increased interest in the chromatographic separation of enantiomeric molecules. However, the resolution of enantiomeric compounds is not an easy task as these molecules have identical physical properties, boiling and melting points, solubility, etc. Thus, enantiomeric compounds cannot be resolved using the usual physical and chromatographic techniques.

The resolution of enantiomeric compounds usually involves their conversion into diastereomers each having different physical properties and which can be easily separated. Using this approach, enantiomers can be resolved through the formation of diastereomeric salts, by the synthesis of a diasteromeric derivative or through the formation of transient diastereomeric complexes.

CHIRAL STATIONARY PHASES (CSP s) FOR HIGH PERFORMANCE LIQUID CHROMATOGRAPHY (HPLC)

The formation of transient diasteromaric complexes provides the basic mechanism of HPLC chiral stationary phases (HPLC-CSPs) where such complexes are formed between the enantiomeric solute molecules and an optically pure molecule or polymer, the chiral selector, bound to the chromatographic stationary phase.

A diagram of a stereochemical resolution on an HPLC-CSP is presented in Fig. 1. Before injection onto the column, the internal energies of the two enantiomeric solutes, *d*-solute and *l*-solute, are identical. The same is true after their elution from the column. In order to separate the two solutes, a difference in internal energy must be produced on the column. This is accomplished through the formation of transient diastereomeric complexes between the solutes and the chiral selector bound to the HPLC-CSP. In the illustration, the solute-CSP complex formed between the *l*-solute and the CSP has a greater energy content, i.e. is less stable, than the solute-CSP complex formed by the *d*-solute. The difference in energies is reflected in the retention times and the greater this difference, the larger the stereochemical resolution.

CLASSIFYING HPLC CSP's

In 1981, the first commercially available HLPC-CSP which operated on these principles was introduced by Pirkle[2]. At that time, the chromatographer was presented with the situation of adapting the the analysis to meet the requirements of the only available HPLC-CSP. However, currently there are 35 commercially available HPLC-CSPs, (Table 1) enabling the chromatographer to select an HPLC-CSP to fit the analysis. The problem is now selecting the right column for a particular separation.

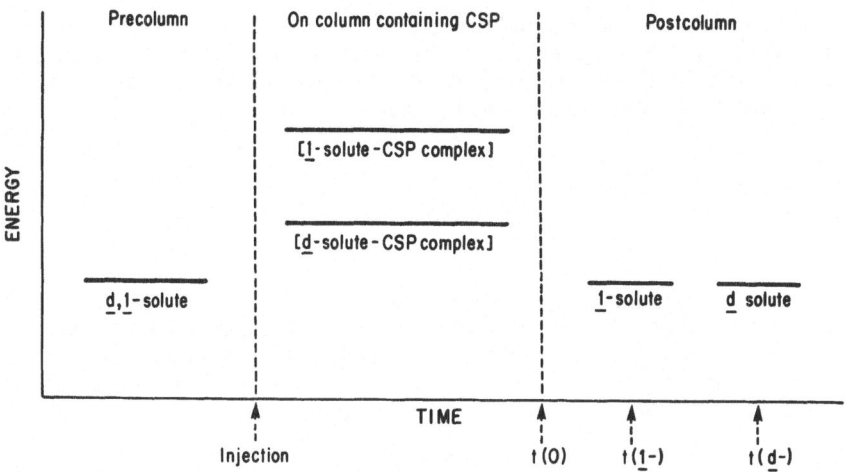

Fig. 1.   General chiral recognition mechanism on a chiral stationary phase.   (Reprinted with permission from J.T. Baker Chemical Company).

A possible solution to this problem is our proposed method for the classification of HPLC-CSP's[3]. In this approach, the phases are divided into 5 classes according to the chiral recognition process which operates on the CSP. This involves the division of this process into its component parts. This can be done if one remembers that the components are interdependent and cannot exist in isolation from one another. Thus the process can be broken down into two main parts:

(1)   How the solute-CSP diastereomeric complexes are formed;
(2)   How the stereochemical differences are expressed during and after the formation of these complexes.

The initial part of this process, the formation of the solute-CSP complexes, can be used as the basis for the division of the current commercially available HPLC-CSPs into common categories as follows.

Type I   Where the solute-CSP complexes are formed by attractive interactions such as hydrogen bonding, pi-pi interactions and dipole stacking. These are the CSPs described by Pirkle[2], and others (CSP's 1-15, Table 1).

Type II   Where the solute-CSP complexes are formed by attractive interactions and through the inclusion of the solute in a chiral cavity or ravine on the CSP. These are the cellulose-based CSPs described by Ichida, et al,[4] and Shibata, et al,[5] (CSP's 16-20).

Type III   Where the primary mechanism involves the formation of inclusion complexes. These are the phenylmethacrylate polymers described by Blaschke[6] and Okamoto and Hatada[7] (CSP 21); the microcrystal-line cellulose triacetate CSP described by Hesse and Hagel[8] (CSP 22); and the cyclodextrin-based CSPs described by Armstrong and De Mond[9] (CSP 23, 24).

Type IV   Where the solute is part of a diastereomeric metal complex which uses an amino acid bound to the stationary phase as one of the ligands. This is chiral ligand exchange chromatography which has been described by Lindner and Pettersson[10] (CSP 25-28).

Type V   Where the CSP is a protein and the solute-CSP complexes are based upon combinations of hydrophobic and polar interactions. This is the CSP based upon immobilized bovine serum albumin described by Allenmark et al.[11] (CSP 29) and the CSP derived from α1-acid glycoprotein developed by Hermansson[12] (CSP 30).

CHOOSING A CSP

Each of the HPLC-CSP classes has a number of distinct characteristics including the structural features of the solutes and the type of mobile phase. Some of these features have been previously addressed [3,13] including a proposed guide for choosing the appropriate type of HPLC-CSP for the resolution of a chiral molecule containing an amine moiety[3]. This is expanded here to simplify the choice of an HPLC-CSP for the resolution of amino acids, chiral compounds containing a carboxylic acid moiety and compounds containing a chiral sulfur atom.

CHOOSING AND HPLC-CSP FOR THE RESOLUTION OF AMINO ACIDS

A guide to the chiral resolution of amino acids is presented in Fig. 2. Racemic amino acids can be stereochemically resolved on Type I, III, IV and V CSPs. There have been no reported resolutions on Type II CSPs. The variety of HPLC-CSPs provides the analyst with the ability to choose a column to fit the analytical needs rather than fitting the assay to the column. The question is now whether to derivatize the solute and what type of mobile phase would be best.

If the decision is to use derivatized solutes, then a Type I, III or V CSP should be chosen. These CSPs require the derivatization of the solute at the amine function and/or the carboxyl moiety. If, however, underivatized solutes are going to be chromatographed, a Type IV CSP is the column of choice.

The same type of choice can be made regarding the mobile phase. If a nonaqueous mobile phase is desired, a Type I CSP can be used. If an aqueous mobile phase is desired, Type III, IV and V CSPs can be used although resolutions of N-3, 5-dinitrobenzoyl aklyl ester derivatives have been reported on Type I CSPs (CSPs 6 and 7) using a mobile phase composed of methanol:water[14].

Enantiomeric and diastereomeric di- and tripeptides can also be stereochemically resolved on HPLC-CSPs. Di- and tripeptides can be resolved on a Type I CPS (CSP 6 and 7) as N-dibenzoyl and O-alkyl ester derivatives using a hexane:2-propanol mobile phase[15]. Cyclic and some linear dipeptides can be resolved on a Type III CSP (CSP 23) without derivatization using a water:methanol mobile phase[16].

CHOOSING AN HPLC-CSP FOR THE RESOLUTION OF CHIRAL MOLECULES CONTAINING A CARBOXYLIC ACID MOIETY

When the major functional group in a chiral molecule is a carboxylic acid moiety, there are a number of possible HPLC-CSPs which can be utilized to stereochemically resolve this type of molecule. These methods are summarized in Fig. 3. When considering a molecule which contains another functional group, for example, an amine moiety, the functionality which is at, or closest to, the chiral center (C*) is the key to the chiral resolution. Therefore, a molecule which has an amine group at the chiral center and a carboxylic acid moiety somewhere else in the molecule will be considered in the resolution of a chiral molecule containing an amine moiety, and vise versa.

Table 1. Commercially available HPLC Chiral Stationary Phases. (Reprinted with permission from J.T. Baker Chemical Company)

Chiral Discriminating Agent	Supplier[a]
CSP Type I	
1 (R)-N-(3,5-dinitrobenzoyl)phenylglycine [covalent][b]	B,R
2 (R)-N-(3,5-dinitrobenzoyl)phenylglycine [ionic][b]	B,R
3 (S)-N-(3,5-dinitrobenzoyl)phenylglycine [covalent][b]	R
4 (S)-N-(3,5-dinitrobenzoyl)leucine [covalent][b]	B,R
5 (S)-N-(3,5-dinitrobenzoyl)leucine [ionic][b]	R
6 D-naphthylalanine	R
7 L-naphthylalanine	R
8 (R)-α-methylbenzylurea	SU
9 (R)-(+)-naphthylethylamine polymer	Y
10 1-(α-naphthyl)ethylamine derivative	SR
11 chlorophenylisovaleroylphenylglycine	SM
12 chrysanthemoylphenylglycine	SM
13 tert.-butylaminocarbonyl valine	SM
14 (S), (S)-α-naphthylethylaminocarbonylvaline	SM
15 (R), (R)-α-naphthylethylaminocarbonylvaline	SM
Type II	
16 cellulose triacetate	B,D
17 cellulose tribenzoate	B,D
18 cellulose trisphenylcarbamate	B,D
19 cellulose tricinnamate	B,D
20 cellulose tris(3,5-dimethylphenylcarbamate)	B,D
Type III	
21 poly(triphenyl methyl methacrylate)	B,D
22 microcrystalline cellulose triacetate	B,O,M,MN
23 β-cyclodextrin	A,SR
24 α–cyclodextrin	A
Type IV	
25 proline	B,D
26 amino acid	B,D
27 hydroxyproline	MN,SR
28 valine	SR
Type V	
29 bovine serum albumin	AN
30 α_1-acid glycoprotein	L

[a]A = A dvanced Separations Technologies, Inc., Whippany, NJ 07981, U.S.A.
AN = Anspec Co., Inc., Ann Arbor, MI 48107, U.S.A.
B = J.T. Baker Chemical Co., Phillipsburg, NJ 08865, U.S.A.
D = Daicel Chemical Industries, Ltd., New York, NY 10166-0130, U.S.A.
L = LKB, Bromma, Sweden
M = E. Merck, Darmstadt, FRG
MN = Macherey, Nagel and Co., Duren, FRG.
R = Regis Chemical Co., Morton Grove, IL 60053, U.S.A.
SM = Sumitoms Chemical, Oska, Japan
SR = Serva, Heidelberg, FRG
SU = Supelco, Inc., Bellefonte, PA 16823, U.S.A.
Y = YMC Inc., Mt. Freedom, NJ 07090, U.S.A.
[b]Type of binding to silica support

	TYPE OF CSP				
	I	II	III	IV	V
Amine Moiety	3,5-DNP[1,2] acetyl[1]	———	undervatized dansyl	undervatized	benzoyl naphthoyl p-nitrobenzoyl 2,4-dinitrobenzoyl
Carboxyl Moiety	alkyl ester[1,2] underivatized[2]	———	β-naphthylamide β-naphthyl ester	undervatized	underivatized
Mobile Phase	hexane-2-propanol[1] methanol:H$_2$O[2]	———	water:methanol	0.25 M copper (II) sulfate	phosphate buffer: 2-propanol

NOTES:

Type I
1. CSP 11, CSP 13, CSP 14
2. CSP 6, CSP 7
3. β-amino acids resolved on CSP 6 as N-3,5-dinitrobenzoyl alkyl ester derivatives with a hexane propanol mobile phase [29]
4. Di- and tripeptides resolved on CSP 6 and CSP 7 as N-dibenzoyl and O-alkyl ester derivatives [15]

Type II
1. No reported resolutions on Type II CSPs

Type III
1. CSP 23, no reported resolutions on other Type III CSPs
2. Best results when either amine or carboxyl moiety derivatized but not both
3. Cyclic dipeptides and some linear dipeptides resolved without dervatization using water:methanol mobile phase [16]

Type IV
1. Some derivatized amino acids can be resolved

Type V
1. CSP 29, no reported resolutions on CSP 30
2. Pi-acid derivatives give better results

REFERENCES:

Type I   14, 15, 27, 28, 29

Type III   9, 16

Type IV   10, 30, 31

Type V   32, 33

Fig. 2.   Choosing an HPLC chiral stationary phase for the resolution of amino acids. (Reprinted with permission from the J.T. Baker Chemical Company).

In all the reported stereochemical resolutions of chiral molecules containing a carboxylic acid moiety, this moiety has been attached to the C*. This appears to be the major structural requirement. Steric bulk and/or an aromatic group at the C* appear to play a role in determining the magnitude of the stereochemical resolution (α). For example, α usually increases as the steric bulk at C* increases.

When Type I and II CSPs are used, the carboxyl group must be converted to a neutral function. The most common derivatization involves the conversion of the acid into an amide. However, some esters have been resolved on these CSPs. Nonaqueous mobile phases, usually haxane modified with 2-propanol, are used with these CSPs.

Type II, IV and V CSPs are usually able to resolve underivatized carboxylic acids. Type III CSPs, however, may require some precolumn derivatization. For example, CSP 22 often requires derivatization of the carboxyl moiety to α methyl ester[17] while CSP 21 will only resolve benzyl esters or benzamides[18]. Aqueous mobile phases and mobile phases composed of methanol are used with these CSPs.

TYPE OF CSP

	I	II
Carboxyl Moiety	3,5-dinitroanalide[1] naphthalenemethylamide[2] benzamides[2] some esters[3,4,5]	benzamide tropic ester[1]
Molecular Structure	carboxyl moiety at C* steric bulk at C* increases α phenyl group at C* increases α	carboxyl moiety at C* steric bulk at C* increases α aromatic moiety at C* increases α 1,2 dicarboxylic acids resolved as benzamides
Mobile phase	hexane:dichloromelthane:ethanol[1] hexane:2-propanol[2]	hexane:2-propanol

NOTES:

Type I   1. CSP 11, CSP 13, CSP 14
              2. CSP 1
              3. Methyl ester of a tetrahydropyrano [3,4-b] indole ring system resolved on CSP 1. CSP-solute interaction probably between indole ring and CSP, [35]
              4. N-aryl-α-amino and 2-carboalkoxy indoline esters resolved on CSP 1. CSP solute interaction probably between nitrogen bearing moieties and CSP [36]
              5. Methyl esters of hydroxy polyenoic C18-fatty acids resolved on CSP 1 [37]

Type II   1. Atropine and homatrapine resolved on CSP 20 [5]

Type III   1. CSP 23
              2. CSP 22 usually requires derivatization of the carboxyl moiety. Simple methyl esters are often sufficient [17]
              3. CSP 21 requires the derivatization of the carboxyl moiety to benzyl ester or benzylamides [7, 18]

Type IV   1. Malic acid has been resolved on these CSPs

Type V   1. CSP 30, no reported resolutions on CSP 29
              2. Tertiary and quaternary amines used as modifiers and may act as ion pairing agents [42, 43]
              3. When amide and esters are the functional groups, the molecule must also contain an amine moiety

Fig. 3.   Choosing an HPLC chiral stationary phase for the resolution of chiral molecules containing a carboxylic acid moiety. (Reprinted with permission from J. T. Baker Chemical Company).

CHOOSING AN HPLC CHIRAL STATIONARY PHASE FOR MOLECULES WITH A CHIRAL SULFUR MOEITY

    Sulfur, like carbon, can be a center of asymmetry and a large number of chiral sulfur compounds are known to exist. A number of chiral sulfoxides, sulfoximines, sulfinamides and sulfimines have been resolved on Type I II, III and V HPLC-CSPs. These methods are summarized in Fig. 4. In most cases an aromatic group has been attached to the chiral sulfur atom although dialkylsulfoxides and dialkylsulfinimines have been resolved on the O-Type CSPs[19]. Another exception is the Type V HPLC-CSP (CSP 29) where resolutions have been accomplished when the aromatic moiety is not attached to the chiral sulfur[20]. Non-aqueous mobile phases are used with Type I, II and in this case III HPLC-CSPs while an aqueous mobile phase was used with the Type V HPLC-CSP.

III	IV	V
underivatized[1] ester[1,2,3] amide[2,3]	underivatized	underivatized[1] amides[2] esters[2]
carboxyl moiety at C* aromatic moiety must be at C*[1,2] steric bulk at C* increases α α-hydroxycarboxylic acids can be resolved[1,2] cyclic dicarboxylic acids can be resolved[3]	carboxyl moiety at C* α-hydroyl group at C* dicarboxylic acids with one α- hydroxyl group can be resolved	carboxyl moiety at C* bulky groups at or near C* or C* part of cyclic structure
phosphate buffer:acetonitrile[1] ethanol:water[2] methanol[3]	0.25 M cooper (II) sulfate	phosphate buffer with modifiers

REFERENCES:

Type I	22, 27, 28, 34, 35, 36, 37
Type II	5, 38, 39
Type III	7, 17, 18, 40, 41
Type IV	10
Type V	22, 23, 42, 43

Fig. 3. (continued)

THE RESOLUTION OF IBUPROFEN - HOW TO BEGIN

The stereochemical resolution of Ibuprogen, Fig. 5 and other α-methylarylacetic acid antiinflammatory agents has received a great deal of attention. This is due to the fact that the (S)-enantiomer of ibuprofen is the active isomer and that the inactive form of the drug, (R)-ibuprofen, is converted in vivo into the active isomer[21].

Since ibuprofen has both a carboxylic acid group and a phenyl ring at the C*, it is likely that the compounds could be resolved on a Type I, II, III or V HPLC-CSP, Fig. 3. Ibuprofen has in fact been resolved as its naphthalenemethyl amide derivative on a Type I HPLC-CSP (CSP 1)[22], on a Type II HPLC-CSP (CSP 17) [I.W. Wainer, unpublished work] and without derivatization on a Type V HPLC-CSP (CSP 30)[23]. Since the solute can be resovled with or without derivatization, the analyst can choose the HPLC-CSP and analytical method to fit the problem. This is illustrated below.

Ibuprofen has a weak chromophore and the UV detection of nanogram quantities is difficult. This will not present a problem to regulatory and quality control chemist where relatively large amounts of the solute are available for analysis. In this case, the method of choice would be the use of the CSP 30 which will allow for automation of the method a rapid throughput.

However, for the pharmacologist and clinical chemist, the problem is the detection of low levels of ibuprofen in biological samples. In this case, precolumn derivatization which adds a moiety with a large UV extinction coefficient or a fluorescent label is the desired approach. Thus, CSP 1 would be the choice. The time lost by the extra steps - isolation and derivatization - will be compensated by an increased in sensitivity.

	TYPE OF CSP				
	I	II	III	IV	V
Type of Sulfur	sulfoxide[1]	sulfoxide sulfoximine sulfinamide sulfilimine	sulfilimine	———	sulfoxide sulfoximine
				———	
Molecular Structure	aromatic moiety at the S*	aromatic moiety at or α to the S*[1]	aromatic moiety at the S*	———	aromatic moiety in molecule
Mobile Phase	hexane:2-propanol	hexane:2-propanol	toluene:dioxane	———	phosphate buffer

NOTES:

Type I 1. CSP 1, CSP 2
2. Spiro-2,2-dithiolane -1-oxides and 1-aryl-1-alkyl-2,2-dithiolane -1-oxides can be resolved on CSP 2 [2].
3. A phenyl ring is the most common substituent but a compound with an α-anthranyl moiety at the S* has been resolved on CSP 2 [44].

Type II 1. Aromatic group at the S* increases stereochemical resolution, however, dialkylsulfoxides and dialkylsulfinimines can be resolved [19, 45].

Type III 1. CSP 21, no reported resolutions on other Type III CSPs

Type IV 1. No reported resolutions on Type IV CSPs

Type V 1. CSP 29, no reported resolutions on CSP 30.
2. The aromatic moiety does not have to be at the S*.
3. A series of 2-pyridinylmethyl-2-benzimidazolyl sulfoxides have been resolved [20].

REFERENCES:

Type I   2, 44

Type II   4, 5, 19, 45

Type III 6

Type V   20, 46

Fig. 4.   Choosing an HPLC chiral stationary phase for the resolution of molecules containing a chiral sulfur atom. (Reprinted with permission from the J. T. Baker Chemical Company).

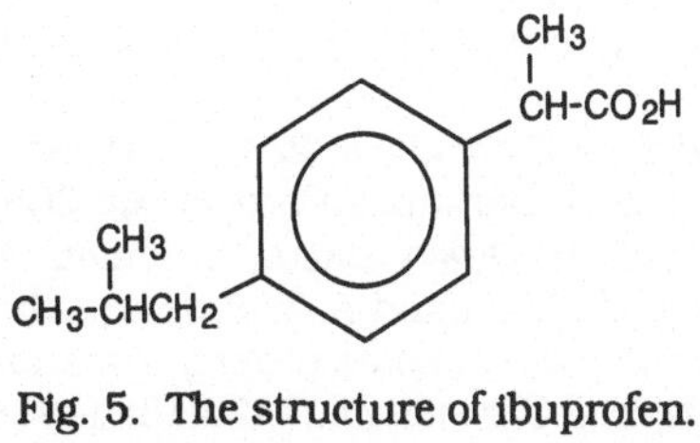

Fig. 5. The structure of ibuprofen.

Different methods for the resolution of the enantiomers of a compound also present the analyst with a choice of mobile phases. For example, the resolution of ibuprofen on the Type I HPLC-CSP is accomplished using a non-aqueous mobile phase composed of hexane: 2-

propanol[22] while the resolution on the Type V HPLC-CSP uses an aqueous mobile phase containing a phosphate buffer and N, N-dimethylethylamine[23].

There are a variety of reasons for choosing a nonaqueous mobile phase. One of these is the coupling of the LC system to a mass spectrometer. This was accomplished by Crowther, et al,[24] for the detection of the enantiomers of ibuprofen in the urine of adult mares. Aqueous mobile phases, on the other hand, offer another set of possibilities including direct analysis of biological samples using coupled-column, or column switching, techniques. The coupling of achiral phase chromatographic systems has been reported by Edholm, et al,[25] and Wainer and Stiffin[26]. the problem of detecting low amounts of an underivatized solute can be also solved in a system using an aqueous mobile phase through coupling to a mass spectrometer[25] or through post column derivatization.

REFERENCES

1. K. M. Giacomini, W. L. Nelson, R. A. Pershe, L. Valdivieso, K. Turner-Tamiyasu and T. F. Blaschke, *J. Pharmacokin. Biopharm* 14:335 (1986).
2. W. H. Pirkle, J. M. Finn, J. L. Schreiner and B. C. Hamper, *J.Am.Chem.Soc* 103:3964 (1981).
3. I. W. Wainer, *Trends in Anal.Chem.* 6:125 (1987).
4. A. Ichida, T. Shabata, I. Okamoto, Y. Yuki, H. Namikashi and Y. Toga, *Chromatographia* 19:280 (1984).
5. T. Shibata, I. Okamoto and K. Ishii, *J.Liq.Chromatogr.* 9:313 (1986).
6. G. Blashke, *J.Liq.Chromatogr.* 9:341 (1986).
7. Y. Okamoto and K. Hatada, *J.Liq.Chromatogr.* 9:369 (1986).
8. G. Hesse and R. Hagel, *Chromatographia* 9:62 (1976).
9. D. W. Armstrong and W. DeMond *J.Chromatogr.Sci* 22:411 (1984).
10. W. Linder and C. Pettersson, *in*: "Liquid Chromatography in Pharmaceutical Development: An Introduction", I. W. Wainer, ed., Aster Publishing Corp., Springfield, Or pp. 63-131 (1985).
11. S. Allenmark, B. Bomgren and H. Boren, *J.Chromatogr.* 264:63 (1983).
12. J. Hermansson, *J.Chromatogr.* 269:71 (1983).
13. I. W. Wainer and M.C. Alemak, *in*: "Chromatographic Chiral Separations", L. Crane and M. Zief, eds., Marcel Dekker, Inc., New York, NY, pp 355-384 (1988).
14. W. H. Pirkle and M. H. Hyun, *J.Chromatogr.* 322:287 (1985).
15. W. H. Pirkle, D. M. Alessi, M. H. Hyun and T. C. Pochapsky, *J.Chromatogr.* 389:203 (1987).
16. J. Florance, A. Galles, Z. Kosarych, K. Langer and C. Martucci, *J.Chromatogr.* 414:313 (1987).
17. E. Francotte, H. Stierlin and T. W. Faigle, *J.Chromatogr.* 346:321 (1985).
18. Y. Okamoto, S. Honda, I. Okamoto and H. Yuki, *J.Am.Chem.Soc.* 102:6971 (1981).
19. Diacel Chemical Industries, Ltd., *Bulletin on Chiracel Columns.* Los Angeles, CA, (1985).
20. S. Allenmark, J.Liq.Chromatogr. 9:425 (1986).
21. D. G. Kaiser, G. J. Vangiessen, R. J. Resicher and W. J. Wechter, *J.Pharm.Sci.* 65:269 (1976).
22. I. W. Wainer and T. D. Doyle, *J.Chromatogr.* 284:117 (1984).
23. G. Schill, I. W. Wainer and S. A. Barkan, *J.Chromatogr.* 365:73 (1986).
24. J. B. Crowther, T. R. Covey, E. H. Dewey and J. D. Henion, *Anal.Chem.* 56:2921 (1984).
25. L. E. Edholm, C. Lindberg, J. Plson and A. Walhagen, *Abstracts Eleventh International Symposium on column Liquid Chromatography*, Amsterdam, June, (1987).
26. I. W. Wainer and R. M. Stiffin *J.Chromatogr.*, 424:158 (1988).
27. N. Oi, M. Nagase, Y. Inda and T. Doi, *J.Chromatogr.* 265:111 (1983).
28. N. Oi and K. Kithara, *J.Chromatogr.* 285:198 (1984).
29. O. W. Griffith, E. B. Campbell, W. H. Pirkle, A. Tsipouras and M. H. Hyun, *J.Chromatogr.* 362:345 (1986).
30. Daicel Chemical Industries Ltd., *Chiralpak, Chiralcel, Technical Brochure, No. 2*, Los Angeles, CA (1986).

31. Daicel Chemical Industries, Ltd., *Chiralpak, Chiralcel, Technical Brochure, No. 4*, Los Angeles, CA. (1987).
32. S. Allenmark, B. Bomgren and H. Boren, *J.Chromatogr.* 316:617 (1984).
33. S. Allenmark and S. Anderson, *J.Chromatogr.* 351:231 (1986).
34. I. W. Wainer and M. C. Alembik, *J.Chromatogr.* 367:59 (1986).
35. C. A. Demerson, L. G. Humber, B. A. Abraham, G. Schilling, R. R. Martel and C. Pace-Asciak, *J.Med.Chem.* 26:1778 (1983).
36. W. H. Pirkle, T. C. Pochapsky, G. S. Mahler and R. E. Field, *J.Chromatogr.* 348:89 (1985).
37. H. Kuhn. R. Wiesner, V. Z. Lankin, A. Nekrasov, L. Alder and T. Schewe, *Anal. Biochem.* 160:24 (1987).
38. K. G. Feitsma, B. F. H. Dreuth and R. A. DeZeeuw, *J.Chromatogr.* 387:447 (1987).
39. A. Mannschreck, H. Koller and R. Wernicke, *Microcrystalline Cellulose Triacetate, A Versatile Stationary Phase for the Separation of Enantiomers.* Kontakte (1), E. Merck, Darmstadt, F.R.G., pp. 40-48, (1985).
40. I. W. Wainer and M. C. Alembik, *J.Chromatogr.* 358:85 (1986).
41. Y. Okamoto, M. Kawashima and K. Hatada, *J.Chromatogr.* 363:173 (1986).
42. J. Hermansson, *J.Chromatogr.* 298:67 (1984).
43. J. Hermansson and M. Eriksson, *J.Liq.Chromatogr.* 9:621 (1986).
44. W. H. Pirkle and J. M. Finn, *J.Org.Chem.* 47:4037 (1982).
45. I. W. Wainer, M. C. Alembik and C. R. Johnson, *J.Chromotogr* 361:374 (1986).
46. S. Allenmark and B. Bomgren, *J.Chromatogr.* 252:297 (1982).

SYSTEMATIC STUDIES OF CHIRAL RECOGNITION MECHANISMS

William H. Pirkle, Thomas C. Pochapsky,
John A. Burke III and Kris C. Deming

School of Chemical Sciences, University of Illinois at
Urbana-Champaign, Urbana, IL 61801, USA

SUMMARY

A general discussion of chiral stationary phase structure/activity relationships is presented. Against this background, a discussion of the systematic approach used in our development of broad-spectrum chiral stationary phases suitable for liquid chromatographic separation of enantiomers is given. Specific applications are cited, and the logic behind the development of N-aryl alanine-derived phases described, their performances compared, and their modes of action addressed.

INTRODUCTION

The chromatographic separation of enantiomers is, in many cases, an almost ideal method for determining enantiomeric purity. Such separations may be accomplished by either gas or liquid chromatography, the latter having the greater potential for preparative as well as analytical separations. Such separations require the intervention of some chiral nonracemic agent, sometimes added to the mobile phase but more often present as the stationary phase. The preparation of broad-spectrum chiral stationary phases (CSPs) for liquid chromatographic separation of enantiomers is discussed here.

The past ten years have produced a heightened interest in CSPs as evidenced by the number of recent reviews of the subject[1-8]. A great many of the CSPs reported, including some which are commercially available, have been developed empirically, but their modes of action are so poorly understood as to usually prevent confident prediction as to a) whether a given pair of enantiomers can be separated by a particular CSP and b) if they are separable, in what order the enantiomers will elute. Often this means that finding a separation method for a given pair of enantiomers will be by trial and error.

Most of these "mechanistically vague" CSPs are derived from biopolymers such as proteins, starches, cellulose, or other polysaccharides. Despite the uncertainty as to when and how they effect enantiomer separation (plus some objections on purely chromatographic grounds) these CSPs are often useful and show relatively broad, if undefined, scope. They often afford enantiomer separations not matched by synthetic CSPs.

Owing to our long-standing interest in chiral recognition mechanisms, our approach to CSP design has taken a different path. Since chiral recognition requires at least three simultaneous interactions between the species involved (at least one of which is stereochemically dependent), we have utilized rather simple chiral compounds having multiple sites (i.e., functionality) from which specifiable types of intermolecular interactions may occur with suitably functionalized "partner" molecules (Fig. 1). The idea is an old one which, rather surprisingly, has not often been implemented[9].

One of our early classes of CSP, derived from N-(3,5-dinitrobenzoyl) α-amino acids, is shown in generalized form in Fig. 2. The dinitrobenzoylamino acid is attached by either an ionic or covalent linkage to an "undercoat" of aminopropyl silane previously bonded to microparticulate silica. While this type of CSP can be modified, the phases derived from phenylglycine or leucine are generally useful. The covalent (Y = NH) phases generally have a longer lifetime, sometimes lasting for years even under heavy usage. A variety of applications for these phases have been reported and detailed chiral recognition mechanisms have been postulated for a variety of analyte classes[10-19]. These mechanisms will not be recounted here, save to point out that conformational rigidity is important and that π–π interactions, hydrogen bonds, steric and dipole-dipole interactions are used in different combinations to achieve separation of suitably constituted analyte enantiomers. Three sets of combinations seem common and, using three basic chiral recognition mechanisms, one can account for the origin and sense (i.e., the elution order) of thousands of enantiomer separations. Each set of interactions entails the approach of the analyte enantiomers to the most accessible face of the immobilized chiral selector. In the more retained enantiomer, the complimentary sites are arrayed more favorably for interaction than in the less retained enantiomer. Because of the variety of complimentary sites available and the number of permutations of site combinations possible, the clientele of enantiomeric analytes which may be resolved on the type 1 CSPs is extremely large. Nevertheless, only those analytes meeting certain (and often specifiable) requirements will resolve on these CSPs.

It is clear that a CSP must interact with analyte enantiomers to form diastereomeric adsorbates which differ in stability if enantiomer separation is to occur. This stability difference is exponentially related to the magnitude of α, the separation factor, (the ratio of the corrected retentions of the two enantiomers). Quite small energy differences produce observable separations on chromatographically efficient HPLC systems. While a separation factor of unity means no separation, a value of 1.05 leads to an easily observable separation and requires a difference in free energy of adsorption of only 30 calories at room temperature. Larger energy differences increase α and facilitate preparative-scale resolution of racemates. While α is a ratio of partition coefficients and not influenced by the absolute rates of adsorption and desorption, these rates are important for they have implications for chromatographic behavior (i.e., bandshapes, run duration, etc.). Polymeric CSPs often show evidence of mass transfer problems due to slow kinetics and therefore tend to give broader bands than bonded-phase CSPs. An additional disadvantage of the biopolymeric CSPs is that they are usually available in only one antipodal form. The totally synthetic CSPs (e.g. 1) may be prepared in either antipodal form or as the racemic analogs. This allows either reversal of elution order or collapse of the enantiomers into a single band simply by switching to a stereochemically different column. This can aid in determining which peaks stem from enantiomer separation in a complex mixture. Additionally, linking a racemic and chiral column in series gives a greater dispersion of peaks than does the chiral column alone[20].

The two greatest advantages to the use of small synthetic chiral selectors are: a) the ability to manipulate structure in virtually any desired fashion and b) the reciprocity of chiral recognition noted in such systems. The latter consideration means that one can use one CSP to develop others. Simply stated, if a CSP derived from (+)-A retains (+)-B, then a CSP comprised of immobilized (+)-B should, using the same interactions, selectively retain (+)-A. The two situations are not "mirror images" and the relationship is only approximate. The manner of immobilization, for example, has some bearing upon the efficacy of the chiral recognition process[21,22]. Even so, any enantiomers which separate on a CSP akin to 1 are themselves candidates for incorporation into a reciprocal CSP expected to separate the enantiomers of analytes structurally related (in terms of complimentary sites) to the type 1 selectors. Since the type 1 CSPs can separate the enantiomers of a great many compounds, it is clear that there is a large number of potential reciprocal CSPs and an almost astronomical number of CSP-analyte combinations which will afford observable levels of chiral recognition. This is grist for many mills indeed! It is equally clear that one wishes to prepare only those reciprocal CSPs which either will afford improved scope,

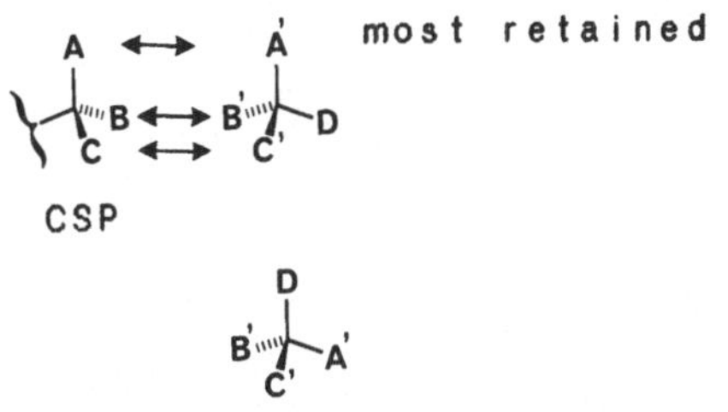

most retained

CSP

Fig. 1.   The three-point rule governing chiral recognition. The CSP has three sites available for attractive interaction (A, B, and C). Only the enantiomer marked "most retained" is capable of simultaneous interaction at all three sites. The less-retained enantiomer can at best maintain two of the three interactions at any one time. Interactions need not be attractive, but at least one must be stereochemically dependent.

efficiency, selectivity or will lead to improved understanding of the requirements of chiral recognition.

This concept of reciprocity of chiral recognition is obviously inapplicable to the design of biopolymer-derived or other polymeric CSPs. Such CSPs, while their mode of action is not yet well understood, do often seem to function as chiral cavities which admit one enantiomer but discriminate against the other. The secondary and tertiary structure of these polymers is profoundly important to their successful functioning as CSPs. It is this inclusion aspect of their behavior that is presumably responsible for their high level of shape selectivity and for the observation that a seemingly innocuous change in analyte structure (e.g., methyl to ethyl) may make the difference between success and failure of the separation. However, this shape selectivity sometimes confers an ability to separate comparatively unfunctionalized enantiomers which cannot be separated on nonpolymeric CSPs.

RESULTS AND DISCUSSION

Among the enantiomers which exhibit large separation factors on type *1* CSPs are the N-aryl-α-amino esters. Systematic studies were conducted on the relationship between analyte structure and the observed enantioselectivity of the CSPs[18,27]. From such relationships, a chiral recognition model was formulated to account for the origin of enantioselectivity, elution order of the enantiomers, and qualitatively, for the extent of enantioselectivity (i.e., the magnitude of α). This model invokes $\pi-\pi$ bonding between the π-basic N-aryl substituent and the π-acidic N-(3,5-dinitrobenzoyl) substituent of the CSP. Two additional non-interchangeable hydrogen bonds were also postulated as was a degree of conformational preference for both the analyte and the CSP. In examining the effect of structure upon chromatographic behavior, it was noted for the enantiomers of ethyl N-arylphenylglycinates that the magnitudes of the separation factors increase in the sequence: phenyl <1-naphthyl <2-naphthyl <2-anthryl. This indicates the π–basicity and conformational behavior of the N-aryl substituent influences the extent of chiral recognition. Increased π–basicity increases both retention and selectivity. The difference between the 1- and 2-naphthyl interaction is rationalized as conformational in origin and stemming from intramolecular interaction of the peri-hydrogen of the 1-naphthyl group with proximate portions of the analyte. Structural optimization led to an N-aryl-α-amino ester the enantiomers of which show a separation factor of ten on a type *1* CSP. This compound, an ester of N-(2-naphthyl)alanine, was an obvious candidate for a reciprocal CSP. While there was some suggestion that an N-(2-anthryl) substituent might afford superior performance, the synthetic ease with which a 2-naphthyl substituent may be appended to the amino nitrogen of an amino acid[28,29] led to the initial preparation of alanine- and valine- derived CSPs *2a* and *2v*[28,29] (Fig. 2).

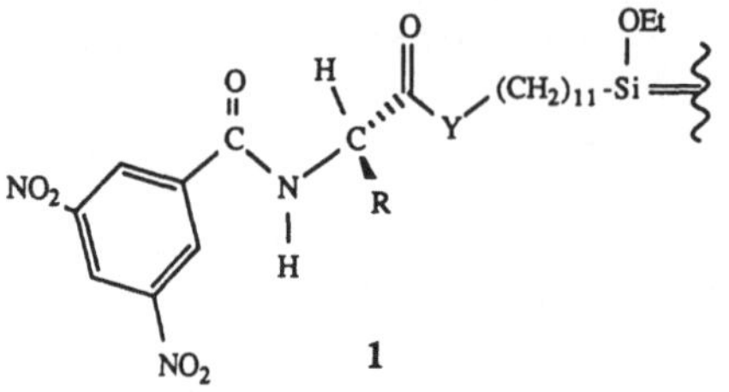

1

R= Phenyl, Isobutyl, Isopropyl, etc.

Y= NH or $-O^- {}^+NH_3-$

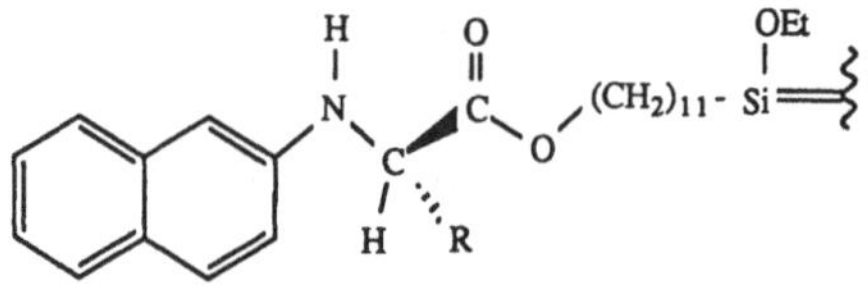

2a R= CH_3

2v R= $CH(CH_3)_2$

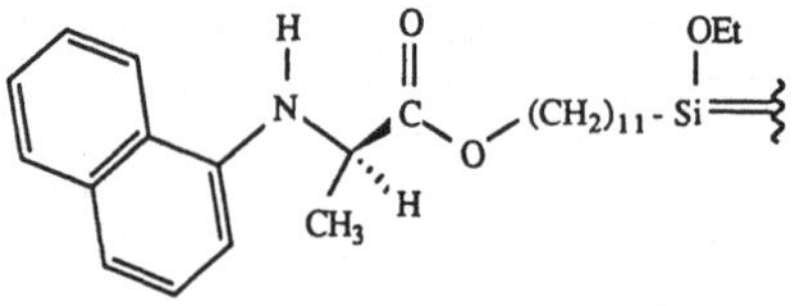

3

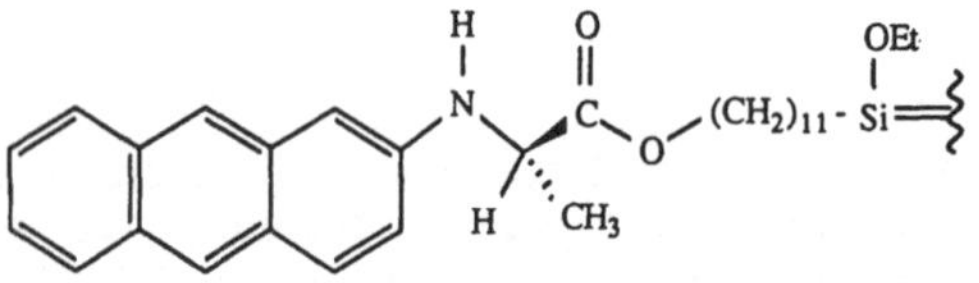

4

Fig. 2. Structure of some CSP s, 1 = N-(3,5-dinitrobenzoyl)-α-amino ester, 2=N-(2-naphthyl)-α-amino esters, 3=N-(1-napthyl)-α-amino ester, 4=N-(2-anthryl)-α–amino ester.

As expected, CSPs *2a* and *2v* separate the enantiomers of N-(3,5-dinitrobenzoyl)-α-amino acid derivatives with great facility. This ability extends to similar derivatives of β– and α-amino acids, α-aminophosphonic acids, and α-arylamino alkanes. Indeed, the enantiomers of the 3,5-dinitrobenzamides of most primary amines having the amino group on the stereogenic center may be separated on these CSPs. The enantiomers of many 3,5-dinitrophenyl ureas have been resolved in similar fashion[31,32].

Using 3,5-dinitrophenylisocyanate as a derivatizing agent, one can convert many chiral secondary alcohols and thiols into derivatives resolvable on CSPs *2a* and *2v*. CSPs *2a* and *2v* usually require derivatization of the analytes since the CSPs were designed to interact with π-acidic sites in the analytes. Such derivatization is simple, uses readily available achiral reagents, does not entail kinetic fractionation of the analyte enantiomers (provided crystallization is avoided) and does not lead to differential detector response toward the derivatives. Derivatization also facilitates detection by providing strongly absorbing ultraviolet chromophores.

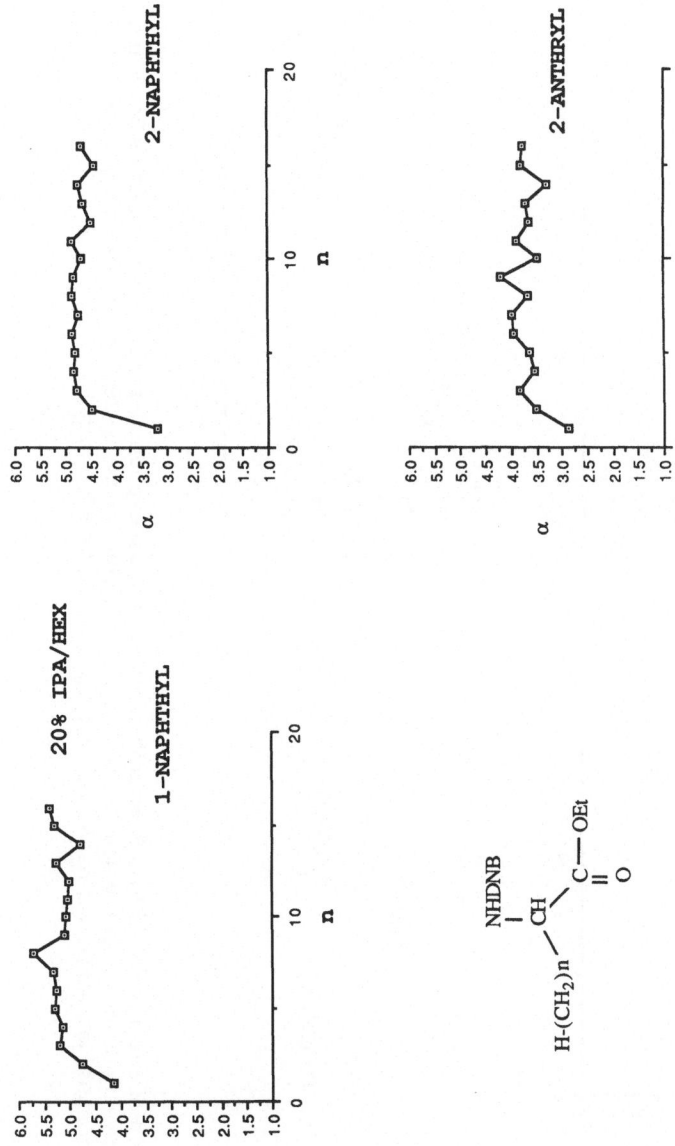

Fig. 3.   Separation of enantiomers of N-(3,5-dinitrobenzoyl)-α-amino esters on CSPs 2a, 3 and 4 as a function of alkyl side-chain length (n). α is the separation factor, the mobile phase was 20% 2-propanol in hexane and the labels refer to which stationary phase is used (2a = 2-naphthyl, 3 = 1-naphthyl and 4 = 2-anthryl, See Fig. 2).

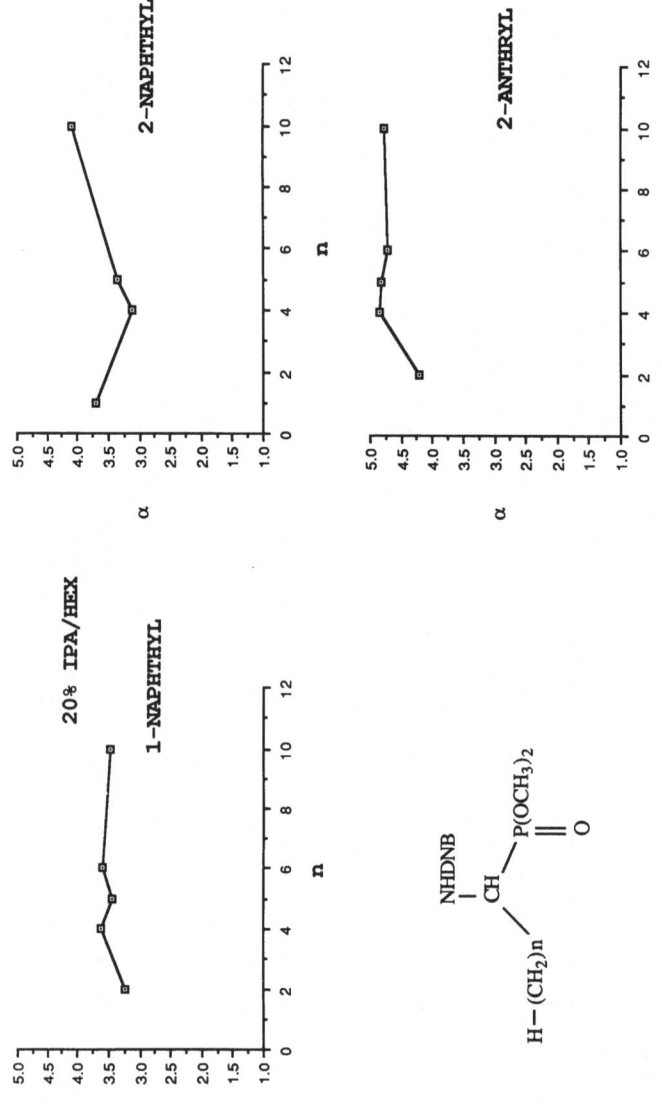

Fig. 4. Separation of enantiomers of N-(3,5-dinitrobenzoyl)-α-aminophosphonate methyl esters on CSPs 2a, 3 and 4 as a function of alkyl side-chain length (n). α is the separation factor, the mobile phase is 20% 2-propanol in hexane and the labels refer to which stationary phase is used (2a = 2-naphthyl, 3 = 1-naphthyl and 4 = 2-anthryl).

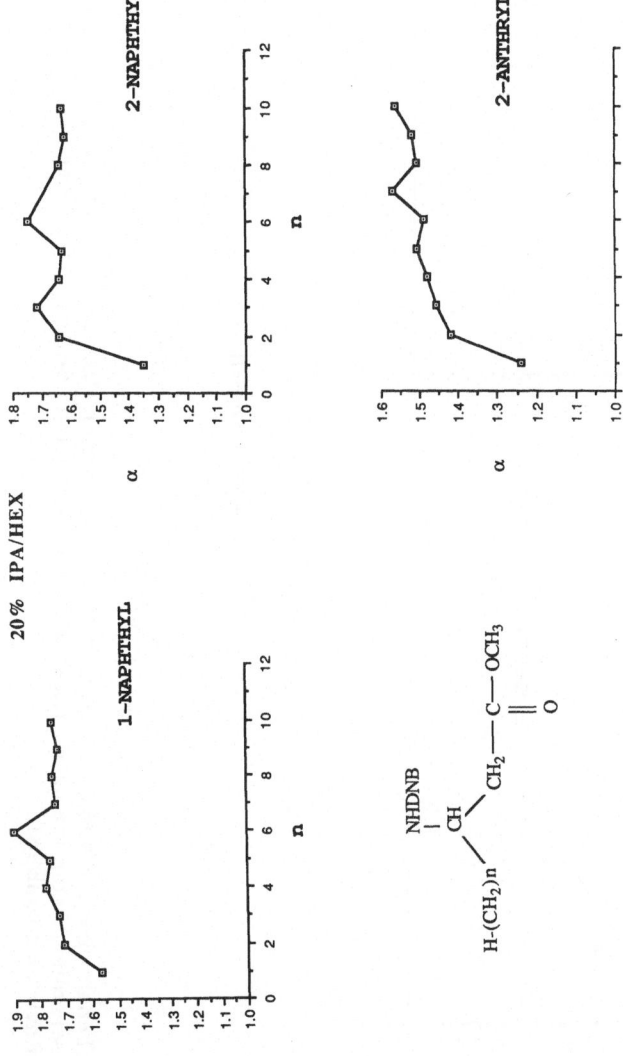

Fig. 5. Separation of enantiomers of N-(3,5-dinitrobenzoyl)-β-amino esters on CSPs 2a, 3 and 4 as a function of alkyl side-chain length (n). α is the separation factor, the mobile phase is 20% 2-propanol in hexane and the labels refer to which stationary phase is used (2a = 2-naphthyl, 3 = 1-naphthyl and 4 = 2-anthryl).

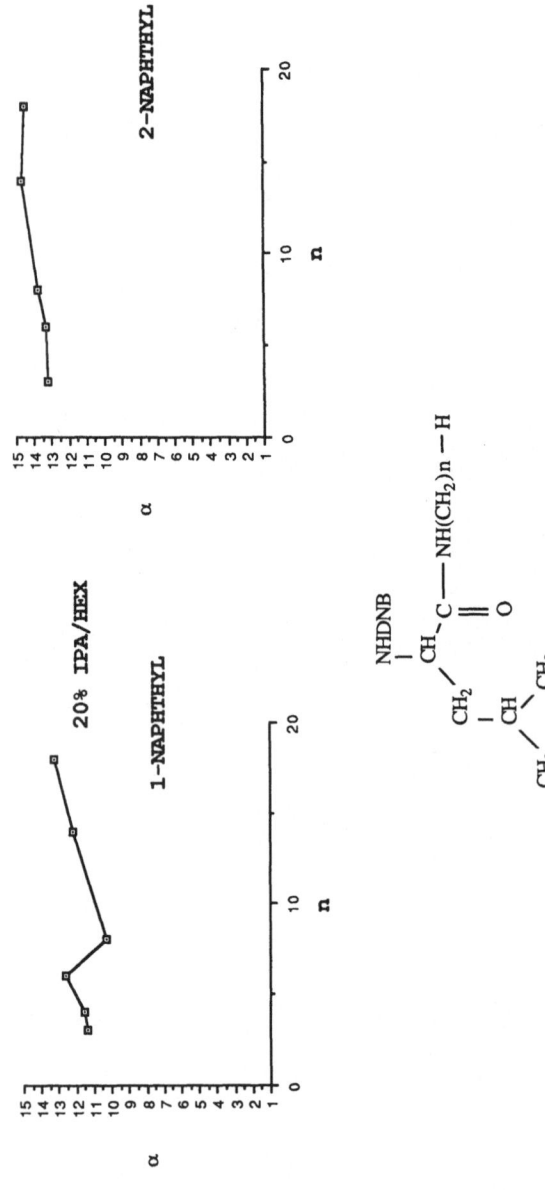

Fig. 6. Separation of enantiomers of N-(3,5-dinitrobenzoyl)leucine amides on CSPs 2a and 3 as a function of secondary amide chain length (n). α is the separation factor, the mobile phase is 20% 2-propanol in hexane and the labels refer to which stationary phase is used (2a = 2-naphthyl, 3 = 2-naphthyl).

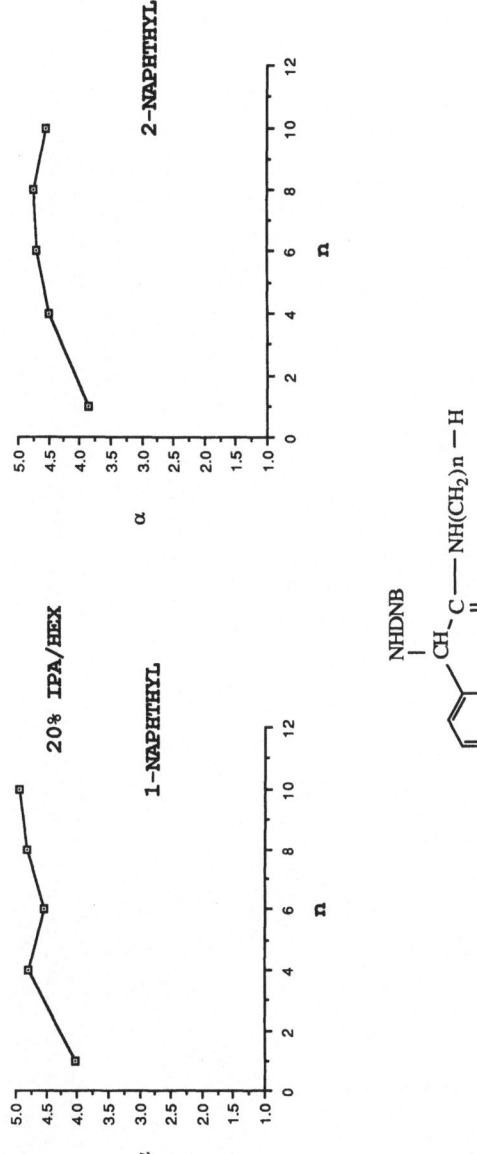

Fig. 7. Separation of enantiomers of N-(3,5-dinitrobenzoyl)phenylglycine amides on CSPs 2a and 3 as a function of secondary amide chain length (n). α is the separation factor, the mobile phase is 20% 2-propanol in hexane and the labels refer to which stationary phase is used (2a = 2-naphthyl, 3 = 2-naphthyl).

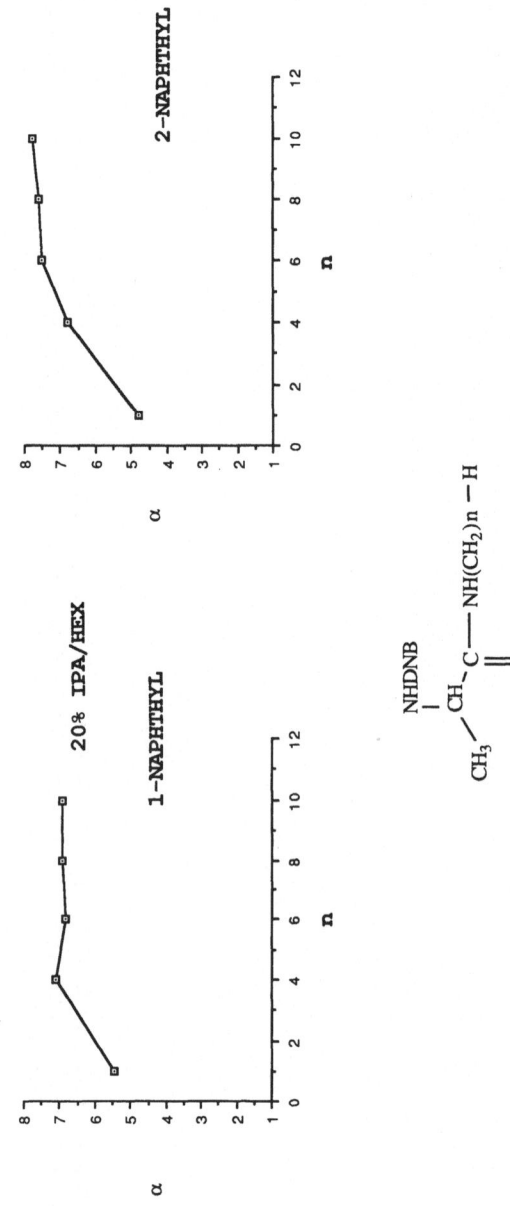

Fig. 8. Separation of enantiomers of N-(3,5-dinitrobenzoyl)alanine amides on CSPs 2a and 3 as a function of secondary amide chain length (n). α is the separation factor, the mobile phase is 20% 2-propanol in hexane and the labels refer to which stationary phase is used (2a = 2-naphthyl, 3 = 2-naphthyl).

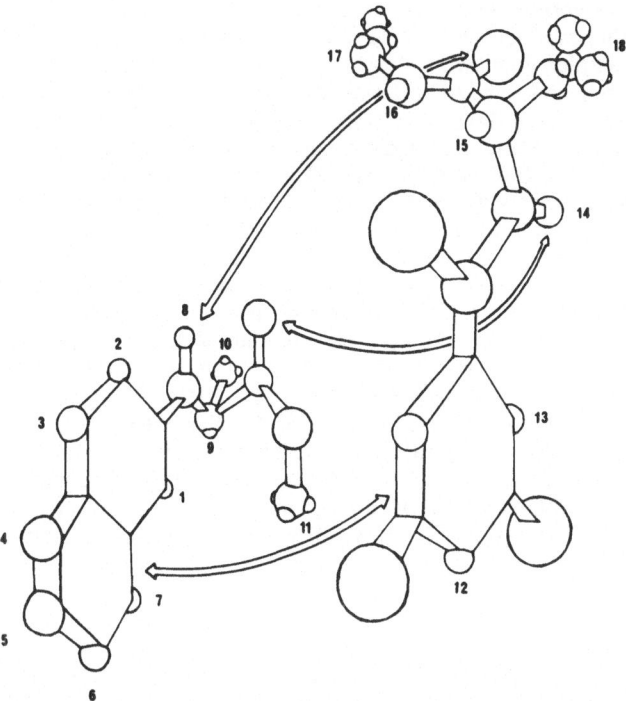

Fig. 9.   Chiral recognition in the more-favored (homochiral) diastereomeric complex
            between soluble analogs of CSP 1 and 2a derived from chromatographic data and
            solution NMR studies, including intra- and intermolecular nuclear Overhauser
            effects (by permission of the American Chemical Society).

Following evaluation of CSPs 2a and 2v, we were interested in exploring the consequences of replacing the N-(2-naphthyl) substituent with other π–donor groups. For the
ethyl phenylglycinates, the N-(1-naphthyl) substituent affords less selectivity than N-(2-
naphthyl) whereas N-(2-anthryl) affords more. CSPs 3 and 4 (Fig. 2) were prepared by
alkylating the aryl amine with ethyl α-bromopropionate, transesterification with 10-
undecenyl alcohol, resolution of the racemic esters on a preparative column packed with
CSP 1, hydrosilylation with trichlorosilane/H_2PtC1_6 and finally, bonding of the chiral
silanes to 5 μM silica. These procedures were similar to those described for 2a and 2v[29,30].
CSPs 3 and 4 have been evaluated using homologous series of analytes, the structures of
which are shown in Figs. 3-8. In each instance, 20% 2-propanol in hexane was used as the
mobile phase in order to facilitate comparison. The choice of mobile phase is relatively
noncritical but does influence retention and, to a lesser extent, selectivity and
chromatographic efficiency (band shape).   The effect of structural variation in a series of N-
(3,5-dinitrobenzoyl) α–amino ethyl esters is shown in Fig. 3.   One can see no profound
difference between CSPs 2a, 3, and 4 although somewhat surprisingly, 3 affords slightly
greater overall selectivity whereas CSP 4 shows slightly lower selectivity. It appears that
structure-activity correlations gathered from phenylglycine derivatives may not carry over
to alanine derivatives. It is unnecessary that they do so, since selectivity stems from CSP-
analyte interaction; if either is altered, the relationships are also changed somewhat. From
Fig. 4, it can be seen that for a series of N-(3,5-dinitrobenzoyl)-α–aminophosphonates CSP 4
now affords the greatest and CSP 3 the least overall selectivity. One will also note from Figs
3-8 that great generality is observed. If one member in the series is resolvable, so are the
others. Selectivity is a function of the class to which the analyte belongs, not of size or any
other single parameter. Elution orders are constant for all members of a given series and are
the same on all three CSPs (of the same absolute configuration). The elution orders are also
consistent with the chiral recognition model formulated for CSP 2a. This model has been
substantiated by a series of spectroscopic studies including the observation of inter-

molecular nuclear Overhauser effects within the more stable diastereomeric complexes formed from soluble analogs of CSP *2a* and a leucine-derived type *1* CSP[33,34]. The structure of this complex is shown in Fig. 9. As shown in Fig. 5, the selectivity for β-amino acid derivatives is reduced relative to the α–analogues, presumably as a consequence of the greater spacing between bonding "sites" in the former and greater conformational mobility. It should be noted that some conformationally less flexible β–amino acid derivatives do show greater selectivities than noted in Fig. 5[35]. Hence the conformations of the analytes do markedly influence selectivity. Figs. 6 and 7 show that greater selectivities are noted for C-terminal amides of α-amino acids than for esters (the C-terminal carbonyl oxygen, a basic site for hydrogen bonding, is more basic in amides than esters) and that α-aryl substituents reduce selectivity relative to α-alkyl substituents. This latter effect stems from a hydrogen bonding interaction with the π-electrons of the aryl group which can occur in the less retained enantiomer. This leads to increased retention of the (still) less retained enantiomer and a consequent reduction of selectivity (the ratio of the two retention parameters, k_1' and k_2'). The modest dependence of selectivity upon the length of the amine portion of the C-terminal amide derivatives is illustrated in Fig. 8. Similar observations (not illustrated) are made for the alcohol component of the C-terminal ester analogues. Indeed, various anilines and amino acids have been used as the amine components of similar C-terminal amide derivatives with no great effects upon overall selectivity. Use of chiral amines leads to diastereomer formation, the type *2*, *3* and *4* CSPs often separating all stereoisomers[36].

CONCLUSIONS

Three N-aryl alanine CSPs have been prepared and evaluated. While the identity of the N-aryl substituent has some quantitative effect upon enantioselectivity, the difference between 1-naphthyl, 2-naphthyl and 2-anthryl substituents is modest and analyte dependent. All three CSPs appear to function by the same basic chiral recognition mechanism, differing only in details too subtle to specify. The three phases do afford different degrees of retention and efficiency and are not of equal synthetic accessability. We conclude that the N-(2-naphthyl) alanine CSP is probably the best suited for the average user, especially since such columns are commercially available (Regis Chemical Co., Morton Grove, IL, USA).

Acknowledgement

This work has been supported by grants from the National Science Foundation and from Eli Lilly and Company.

REFERENCES

1. G. Schill, I. W. Wainer, and S. A. Barkan, *J.Liq.Chromatogr.*, 9:641-666.
2. R. W. Souter, "Chromatographic Separation of Stereoisomers", CRC Press, Cleveland, OH (1985).
3. W. H. Pirkle and J. M. Finn *in*: "Assymetric Synthesis, Analytical Methods", J.D. Morrison, ed., Academic Press, NY, NY Ch.6 (1983).
4. B. C. Hamper, *Science*, in press.
5. W. H. Pirkle and T. C. Pochapsky *in*: "Advances in Chromatography," Vol. 27, Giddings, Brown and Grushka, eds., Marcel Dekker, Inc. 73-127. (1987).
6. T. C. Pochapsky, *Biochromatography*, 2 28-35 (1987).
7. A. Mannschreck, H. Koller, and R. Wernicke, *Kontakte* E. Merck, Darmstadt) 1:40-48.
8. R. Dappen, H. Arm and V. R. Meyer, *J.Chromatogr.*, 373:1-20 (1986).
9. C. E. Dalgleish, *J.Chem.Soc.* 3940-3942 (1952).
10. W. H. Pirkle, D. W. House, and J. M. Finn, *J.Chromatogr.*, 192:143 (1980).
11. W. H. Pirkle and J. M. Finn, *J.Org.Chem.*, 46:2935 (1981).
12. W. H. Pirkle and J. L. Schreiner, *J.Org.Chem.*, 46:4988-4991 (1981).
13. W. H. Pirkle, J. M. Finn, J. L. Schreiner and B. C. Hamper, *J.Am.Chem.Soc.*, 103:3964 (1981).
14. W. H. Pirkle, J. M. Finn, J. L. Schreiner, B. C. Hamper and J. R. Pribish *in*: "Am. Chem. Soc. Symposium Series," No. 185, Eliel and Otsuka, eds., Chp.18, 245 (1982).

15. W. H. Pirkle, C. J. Welch and M. H. Hyun, *J.Org.Chem.*, 48:5022 (1983).
16. W. H. Pirkle and C. J. Welch, *J.Org.Chem.*, 49:138 (1984).
17. W. H. Pirkle, M. R. Robertson and M. H. Hyun, *J.Org.Chem.*, 49:2433 (1984).
18. W. H. Pirkle, T. Ch. Pochapsky, G. S. Mahler and R. E. Field, *J.Chromatogr.*, 348:89-96 (1985).
19. W. H. Pirkle and T. J. Sowin, *J.Chromatogr.*, 387:313-321 (1987).
20. W. H. Pirkle, R. Dappen and D. S. Reno, *J.Chromatogr.*, 407:211-216 (1987).
21. W. H. Pirkle, M. H. Hyun and B. Bank, *J.Chromatogr.*, 316:585-604 (1984).
22. W. H. Pirkle, M. H. Hyun, A. Tsipouras and B. Bank, *J.Pharm.Biomed.Anal.*, 2 173-182 (1985).
23. G. Blaschke, *J.Liq.Chromatogr.*, 9:341 (1986).
24. G. Wulff, H. G. Poll and M. Minarek, *J.Liq.Chromatogr.*, 9:385-406 (1986).
25. T. Shibata, I. Okamoto and K. Ishii, *J.Liq.Chromatogr.*, 9:313-340 (1986).
26. Y. Okamoto and K. Hitada, *J.Liq.Chromatogr.*, 9:369-384 (1986).
27. T. C. Pochapsky, Ph.D. Thesis, University of Illinois at Urbana-Champaign, (1986).
28. W. H. Pirkle and T. C. Pochapsky, *J.Am.Chem.Soc.*, 108:352-354 (1986).
29. W. H. Pirkle and T. C. Pochapsky, *J.Org.Chem.*, 51:102 (1986).
30. W. H. Pirkle, T. C. Pochapsky, G. S. Mahler, D. E. Corey, D. S. Reno and D. M. Allesi, *J.Org.Chem.*, 51:4991 (1986).
31. W. H. Pirkle, G. S. Mahler and M. H. Hyun, *J.Liq.Chromatogr.*, 9:443 (1986).
32. W. H. Pirkle, G. S. Mahler, M. H. Hyun and T. C. Pochapsky, *J.Chromatogr.*, 388:307-314 (1987).
33. W. H. Pirkle and T. C. Pochapsky, *J.Am.Chem.Soc.*, 108:5627 (1986).
34. W. H. Pirkle and T. C. Pochapsky, *J.Am.Chem.Soc.*, 109:5975-5982 (1987).
35. W. H. Pirkle, A. Tsipouras, M. H. Hyun, D. J. Hart and C.-S. Lee, *J.Chromatogr.*, 358:377 (1986).
36. W. H. Pirkle, D. M. Allesi, M. H. Hyun and T. C. Pochapsky, *J.Chromatogr.*, 398:203-209 (1987).

A *NOTE ON* SEPARATION OF ENANTIOMERS OF OXYPHENONIUM BROMIDE BY

HIGH-PERFORMANCE LIQUID CHROMATOGRAPHY

Karla G. Feitsma*, Ben F.H. Drenth and Rokus A. de Zeeuw

Department of Analytical Chemistry and Toxicology
University of Groningen, A. Deusinglaan 2
9713 AW Groningen, The Netherlands

INTRODUCTION

Oxyphenonium bromide (Fig. 1) is a quaternary ammonium compound with strong anticholinergic properties. As with many other anticholinergics, the enantiomers of this drug exhibit large differences in therapeutic effects as well as in biliary and urinary excretion[1]. For a detailed study of the fate of oxyphenonium bromide enantiomers in the body, an assay is required that allows the simultaneous determination of the two enantiomers in the same sample. High performance liquid chromatographic (HPLC)-techniques are preferred, as quaternary ammonium compounds are very sensitive to decompositon upon injection in GC.

Four chiral HPLC-systems were tried for this particular problem:

- a chiral ion-pairing system
- dinitrobenzoylphenylglycine (DNBPG)-bonded stationary phases
- β-cyclodextrin, covalently bonded to silica gel
- an α_1-acid glycoprotein bonded phase (EnantioPac®)

EXPERIMENTAL

Reagents

N-Ethoxycarbonyl-2-ethoxy-1,2-dihydroquinoline (EEDQ), *l*-phenyl-glycine and 3,5-dinitrobenzoylchloride were obtained from Janssen (Beerse, Belgium).

All solvents and compounds used as mobile phase components were of analytical grade and obtained from Merck (Darmstadt, FRG). Racemic oxyphenonium bromide was a gift from Ciba-Geigy (Basle, Switzerland).

Synthesis of Chiral Phases

DNBPG-covalently bonded stationary phase:

(R)-3,5-dinitrobenzoylphenylglycine (DNBPG) was prepared from 3,5-dinitro-benzoylchloride and *l*-phenylglycine. The product was coupled to aminopropyl-silica gel (Spherisorb, 5 μm, Phase Sep, Queensferry, UK) using EEDQ[2]. A column, dimensions 150 x 4.6 mm i.d., was packed with this material using a balanced-density slurry method[3].

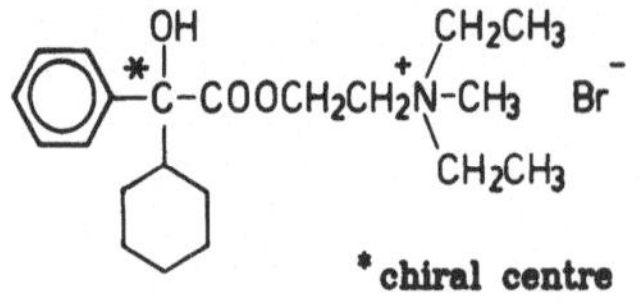

oxyphenonium bromide

*chiral centre

Fig. 1. Structure of oxyphenonium bromide.

DNBPG-ionically bonded stationary phase:

A column, dimensions 250 x 4.6 mm i.d., was packed with aminopropyl-silica gel (Si 100 Polyol, 5 μm, Serva, Heidelberg, FRG). The stationary phase was activated by passing triethylamine in tetrahydrofuran through the system, followed by a solution of 1 g (R)-3,5-dinitrobenzoylphenylglycine in 50 ml tetrahydrofuran. This solution was recycled during 3 hours.

Chromatographic Conditions

-Chiral ion pairing system

Chromatography was performed on a Nucleosil 5 CN (Machery-Nagel, Duren, FRG), column, 150 x 4.6 mm i.d. with a mobile phase of d-camphorsulphonic acid (1×10^{-2}-2×10^{-3} M) in mixtures of heptane, isopropanol and chloroform or dichloromethane at a flow rate of 1 ml/min. Detection was by UV, at various wavelengths, as indicated

- DNBPG-bonded stationary phases

Chromatography was performed on stationary phase of (R)-DNBPG, covalently bonded to aminopropyl-silica gel, (150 x 4.6 mm i.d.) or (R)-DNBPG, ionically bonded to aminopropyl-silica gel, (250 x 4.6 mm i.d) using a mobile phase of mixtures of hexane or dichloromethane with 0 -10% isopropanol at 1ml/min. Detection was by UV. at various wavelengths, as indicated.

- β–Cyclodextrin-bonded phase

Chromatography was performed on a stationary phase of β–cyclodextrin, covalently bonded to silicagel, (175 x 4.6 mm i.d.[4]) with a mobile phase of phosphate buffers with varying amounts of acetonitrile or methanol, different pH's and addition of counter ions like perchlorate, bromide, pentanesulphonic acid and sodium dodecylsulphonate at 1 ml/min. Detection was by UV at 205 nm.

- α_1-Acid glycoprotein stationary phase

Chromatography was performed on an α_1-acid glycoprotein (AGP) column (100 x 4.0 mm i.d.) (EnantioPac®, LKB-AB, Bromma, Sweden), at 24.5°C, and with a mobile phase of aqueous solutions containing 0.02 M sodium phosphate buffer pH = 6.9 and 0.1 M sodium chloride, with 4 - 8% v/v isopropanol delivered at a flow rate 0.2-0.3 ml/min. Detection was by UV at 205 nm and/or diode array (200-350 nm).

Apparatus

The HPLC solvent delivery systems used were a Waters M45 pump, a Perkin Elmer Series 10 pump or a Spectra Physics 3500B pump. The experiments were performed using a Spectra Physics 770 UV detector and/or a HP 8450A (Hewlett-Packard) multi-channel UV-visible diode array spectrophotometer, a Rheodyne 7125 sample injector equipped with a 20 µl loop and a Kipp BD40 recorder. The diode array spectrophotometer was equipped with a model 178.32 QS quartz cell (Hellma, Muhlheim, FRG) of 8 µl. Details of this detector are described elsewhere[5].

RESULTS AND DISCUSSION

Pettersson and Schill separated enantiomeric amines by ion-pair chromatography using d-camphorsulphonic acid as a chiral counter ion[6]. Oxyphenonium, being a quaternary ammonium compound, was also expected to form ion-pairs with this chiral acid. However, we were unable to resolve racemic oxyphenonium using such a system. Pettersson experienced similar problems in resolving the enantiomers of atropine, a related compound, with this system. The lack of separation may be due to the relatively long distance between the amine group and the chiral center with its hydroxyl group, possibly preventing adequate ion-pair formation.

Another approach to separating enantiomers is the use of chiral stationary phases. The phases developed by Pirkle are among the ones most widely applied[7]. Two phases, having (R)-dinitrobenzoylphenylglycine covalently or ionically bonded to silicagel, were synthesized. Bisnaphthol, a test compound, was separated quite well using these stationary phases and non-polar mobile phases; yet, oxyphenonium could not be resolved into its enantiomers. Here too, the distance between the chiral center and the nitrogen may be unsuitable. Another important factor may be that the nitrogen in oxyphenonium is too polar for effective interaction with these chiral phases. Moreover, as was shown for derivatives of tropic acid like atropine, resolution is highly promoted by the presence of an amide bond in the sample molecule[8].

The acid function of oxyphenonium, cyclohexylphenylglycolic acid, could be resolved into enantiomers on β–cyclodextrin-bonded stationary phases[4]. However, here again it was not possible to resolve racemic oxyphenonium itself. Conceivably, the presence of the quaternary nitrogen in the molecule may interfere too much in the separation mechanism.

The development of a chiral stationary phase, consisting of α_1-AGP, covalently bonded to silicagel was described by Hermansson[9]. Recently, Schill et al, reported the chiral separation of cationic drugs using an AGP-bonded chiral phase[10]. Using different mobile phases, compounds like atropine, methylatropine, oxyphencyclimine, methorphan and promethazine were separated into their enantiomers. These results supported the idea that this column might be suitable for the separation of racemic oxyphenonium bromide.

Our attempts to separate a racemate of oxyphenonium bromide into enantiomers using an EnantioPac® column were indeed successful and Fig. 2 shows a three-dimensional chromatogram of racemic oxyphenonium bromide.

It was observed that the amount of organic modifier, i.e. isopropanol, had a large impact on the chromatography (Table 1)): High isopropanol concentrations resulted in short retention times, but this was at the cost of selectivity and resolution.

Some advantages of this chiral stationary phase were detailed by Hermansson et al,[11]: The column can be used in a reversed-phase mode, allowing the regulation of retention and selectivity as well as permitting the direct injection of aqueous samples. In addition there is no need for derivatization, and the column has a long life-time. However, in our opinion, this system also has various disadvantages. Thus although the resolution of racemic oxyphenonium was obtained the flow had to be limited to 0.5 ml/min, and also the amount which could be loaded on to the column was limited (5 nmoles). For compounds with low UV-absorbtion this gives rise to detection problems. Moreover, both the efficiency of the system (N ≅300) and the stability of the baseline were poor (See Fig. 3). For these reasons this chiral system will be of limited use in bioanalysis as well as in the determination of enantiomeric purity of oxyphenonium bromide.

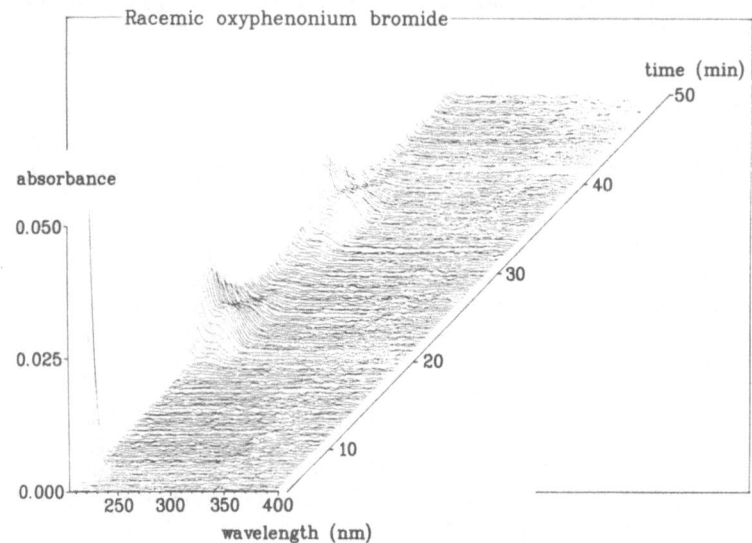

Fig. 2.   Three-dimensional chromatogram of racemic oxyphenonium bromide
stationary phase: EnantioPac®, 100 x 4.0 mm i.d. mobile phase: 5% v/v
isopropanol in 0.02 M sodium phosphate buffer pH-6.9 and 0.1 M sodium
chloride at 0.2 ml/min. detection: UV, at 200-400 nm (diode-array) sample: 2 µg
racemic oxyphenonium bromide flow: 0.2 ml/min.

Table 1.   Impact of Percentage Isopropanol in the Mobile Phase on Selectivity and
Resolution of Oxyphenonium Bromide Enantiomers
Chromatographic conditions as for Fig. 2.
Sample: 2 µg racemic oxyphenonium bromide

isopropanol (% v/v)	$V_{R,l}$[1] (ml)	α[2]	R_s[3]
8	4.6	1.4	1.1
6	7.1	1.6	1.8
5	9.6	1.8	2.2
4	16.1	1.9	2.7

1    $V_{R,l}$ = retention volume of l-oxyphenonium bromide, flow rate 0.2 ml/min
2   α = selectivity (≥1), in this case k'_l/k'_d; k'_l and k'_d being capacity factors of l- and
                    d-oxyphenonium, respectively
3   R_s = resolution, defined as $(\Delta t_R)/2(\sigma_l + \sigma_d)$

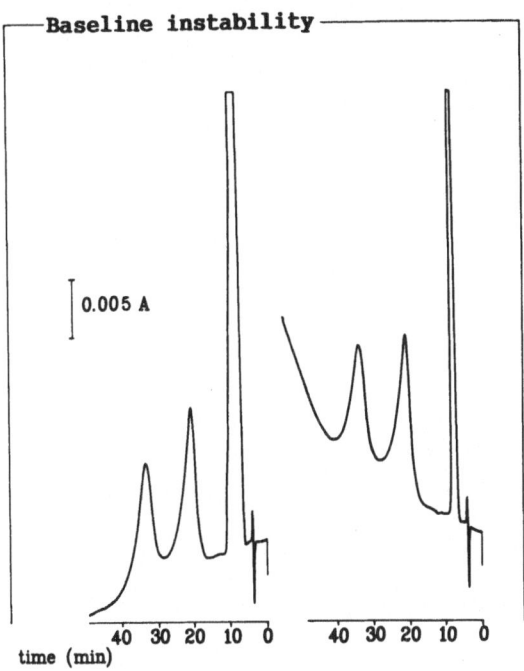

Fig. 3.   Chromatograms of racemic oxyphenonium bromide on EnantioPac®, taken at two consecutive runs. Sample: 2 µg racemic oxyphenonium bromide detection: UV, at 205 nm. Other conditions as for Fig. 2.

In conclusion, many chiral HPLC-systems have been described for the resolution of racemic compounds into enantiomers[12,13]. However, although they may be very useful for selected drugs, it should be observed that these systems are not universal for a given class of compounds and thus, it is not always easy to find a suitable system for a particular separation problem. Moreover, the efficiency of the systems is often insufficient for use in bioanalysis, although applications in this field have béen described (e.g. 14).

REFERENCES

1.   K. G. Feitsma, Ph. D. Thesis, University of Groningen, 1987.
2.   W. H. Pirkle, D. W. House, and J. M. Finn, *J.Chromatogr.*, 192 :143 (1980).
3.   K. Kuwata, M. Uebori and Y. Yamazaki, *J.Chromatogr.*, 211:375 (1981).
4.   K. G. Feitsma, B. F. H. Drenth and R. A. de Zeeuw, *J.Chromatogr.*, 387:447 (1987).
5.   R. T. Ghijsen, B. F. H. Drenth, F. Overzet and R. A. de Zeeuw, *J.HighResolut. Chromatogr.Chromatogr.Commun.*, 5:192 (1982).
6.   C. Pettersson and G. Schill, *J.Chromatogr.*, 204:179 (1981).
7.   R. Däppen, H. Arm and V. R. Meyer, *J.Chromatogr.*, 373:1 (1986).
8.   I. W. Wainer, T. D. Doyle and C. D. Breder, *J.Liq.Chromatogr.*, 7:731 (1984).
9.   J. Hermansson, *J.Chromatogr.*, 298:67 (1984).
10. G. Schill, I. W. Wainer and S. A. Barkan, *J.Liq.Chromatogr.*, 9:641 (1986).
11. J. Hermansson and M. Eriksson, *J.Liq. Chromatogr.*, 9:621 (1986).
12. D. W. Armstrong, *J.Liq.Chromatogr.*, 7:353 (1984).
13. B. Testa, *Xenobiotica*, 16:265 (1986).
14. J. Hermansson and M. Eriksson, *J.Chromatogr.*, 336:321 (1984).

THE USE OF PIRKLE HIGH-PERFORMANCE LIQUID CHROMATOGRAPHY PHASES

IN THE RESOLUTION OF ENANTIOMERS OF POLYCYCLIC AROMATIC

HYDROCARBON METABOLITES

Michael Hall* and Philip L. Grover

Chester Beatty Laboratories
Institute of Cancer Research, Royal Cancer Hospital
Fulham Road, London SW3 6JB, UK

SUMMARY

Polycyclic aromatic hydrocarbons (PAH) are metabolised within tissues in both a regio- and a stereoselective manner. Enantiomers of PAH metabolites have been resolved on high performance liquid chromatography by one of two methods, either (i) indirectly, following derivatization, or (ii) directly using chiral stationary phases (Pirkle columns). This latter technique has been used to assess, for example, the stereoselectivity of benzo(a)pyrene (BP)-7,8-dihydrodiol formation by samples of human skin following the topical application of [3H]BP and the optical purity of BP-7,8-dihydrodiol and anthracene-1,2-dihydrodiol extracted from incubations of [3H]BP and of [14C]anthracene with rabbit skin and liver microsomal fractions.

INTRODUCTION

Polycyclic aromatic hydrocarbons (PAH) are a group of compounds with widespread environmental distribution that are produced by the incomplete combustion of materials of organic origin. They include several compounds known to be potent carcinogens in animal systems[1]. In addition, epidemiological studies have indicated a strong association between an increased incidence of human skin cancer and exposure of an individual to these materials[2]. PAH are metabolised within mammalian tissues to more polar and hence more readily excretable products. However, this metabolism may also yield forms of the parent hydrocarbons, known as ultimate carcinogens, which are capable of binding covalently to DNA, a reaction generally regarded as an initial step in chemical carcinogenesis[3]. The ultimate carcinogen is most commonly a vicinal bay-region diol-epoxide[3,4], a type of metabolite first identified in the case of benzo(a)pyrene (BP) and characterized as the 7,8-dihydro-7,8-dihydroxybenzo(a)pyrene 9,10-oxide[5].

The metabolic activation of PAH via diol-epoxides involves a sequence of three enzyme-catalysed steps: mono-oxygenation by cytochrome P-450, hydration by epoxide hydrolase and further mono-oxygenation again involving cytochrome P-450 (Fig. 1). This is both a regio- and a stereoselective process[4,6] the proportion of parent hydrocarbon finally appearing in the form of its ultimate carcinogen being a reflection of the specificities and relative levels within the tissue of the enzymes involved. In the case of BP, four stereoisomers of the bay-region diol-epoxide may be formed. Of these, the (+)-r-7,t-8-dihydroxy-t-9,10-oxy-7,8,9,10-tetrahydrobenzo(a)pyrene, derived from the proximate carcinogen (-)-trans-7R, 8R-dihydro-7R,8R-dihydroxybenzo(a)pyrene (BP-7R,8R-dihydrodiol) (Fig. 1), has been found to possess greater biological activity than the other three isomers[4,6]. This is also the stereoisomer formed in largest amounts by liver

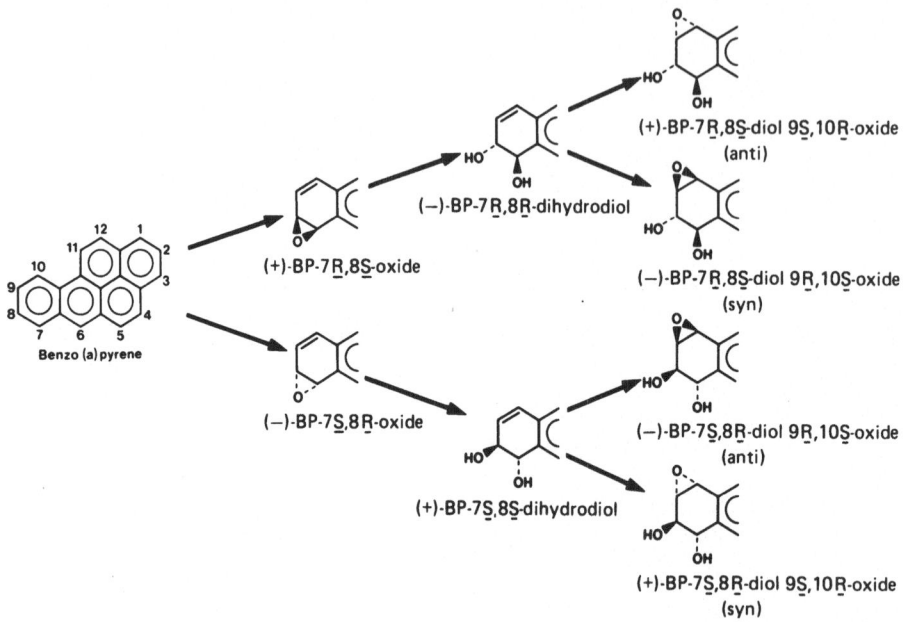

Fig. 1. Stereochemistry of metabolism at positions 7 and 8 of benzo(a)pyrene.

microsomes from 3-methylcholanthrene (3-MC)-treated rats and by purified cytochrome P-450c in a reconstituted system[4,7].

Since the stereoselectivity of PAH metabolism may be an important factor in determining carcinogenic potential, much research effort has been expended on the resolution of enantiomers of different metabolites of various hydrocarbons. This has been achieved mainly by one of two methods. The first involves derivatization with a second compound, itself usually containing a chiral centre, followed by high-performance liquid chromatography (HPLC) on standard normal or reversed-phase columns; the second technique involves direct resolution by HPLC using a chiral stationary phase (CSP) column. A brief synopsis of how these methods have been used and the problems they have been engaged to solve is given below.

The derivatizing agents most commonly employed in the resolution of PAH epoxide and dihydrodiol enantiomers are (-)-menthoxyacetyl chloride and (-)-α-methoxy-α-trifluoromethylphenylacetyl chloride, although there is also one report[8] in which the optical purities of benz(a)anthracene (BA)-5,6- and -8,9-oxides formed by isolated cytochrome P-450c were determined by conjugation with glutathione. Using these two chiral compounds, HPLC separation of metabolite enantiomers of, for example, BP [9-11], BA[12,13], benzo(c)phenanthrene (B(c)Ph)[14-16], naphthalene and anthracene[17,18] has been achieved. This has permitted determination of the absolute configurations of these metabolites[11,14] and has also been used to determine the stereoselectivities of purified epoxide hydrolase[10, 18-20] and of isolated isozymes of cytochrome P-450[12,15-18]. Such studies have led in turn to a proposal for the structure of the binding site of cytochrome P-450c[21]. For this latter type of work derivatization is a particularly suitable technique since the metabolite formed by the enzyme is trapped before it can be either further metabolised or, in the case of epoxides, rearranged to a phenol. There are, however, several disadvantages associated with the procedure, including the possibility of enantiomeric contamination of the derivatizing agent, differential rates of reaction of the two enantiomers with the derivatizing agent, inherent instabilities of the metabolites and low availability of the metabolites[22,23]. These latter points are accentuated when the absolute

44

configurations of resolved enantiomers are required, since further work-up is necessary to remove the derivatizing agent before this can be attempted.

It was for reasons such as these that Yang and his colleagues first applied an HPLC technique for resolving PAH metabolite enantiomers directly[23] employing a CSP of (*R*)-*N*-(3,5-dinitrobenzoyl)phenylglycine ionically bonded to γ–aminopropyl silanised silica, as developed by Pirkle and co-workers[24,25]. The initial report of Weems and Yang[23] described resolution of dihydrodiol enantiomers of BP, BA and 11-methyl BA. Since then refinement of the methodology, together with the commercial availability of covalently-bonded phenylglycine- and ionically- and covalently-bonded leucine-based columns, have led to the direct resolution of the enantiomers of phenol, epoxide and dihydrodiol derivatives of, amongst other PAH, BP[26-29], BA[26-32], B(c)Ph [33,34], anthracene[35] and chrysene [28,32,36]. In addition, the effects of either methyl- or halogen-substitution of the PAH nucleus on both the absolute configurations of the metabolites and also the stereoselectivity of their formation by rat liver microsomes have been investigated. This is perhaps best illustrated in the case of BA, where the effects of monomethyl- [28,31,37-39], dimethyl- [28,31,40-43], and halogen-substitution [31,38,44,45] on these factors have been studied. In general, data on the stereoselectivity of PAH epoxide and dihydrodiol formation by microsomal metabolising systems, and on the absolute configurations of the metabolite enantiomers obtained by this direct method have been found to be qualitatively similar to results obtained using the derivatization technique. However, quantitative differences have been observed in some cases, for example with B(c)Ph-3, 4-oxide and -3,4-dihydrodiol formation[15,16,33,34].

The majority of the studies in which the stereoselective metabolism of PAH has been investigated have been performed using rat liver microsomes as a metabolising system. They have shown that, in the majority of cases, one metabolite enantiomer predominates which, in the case of dihydrodiols, is usually that having *R,R* absolute configuration[4,7]. Adult rat liver is not, however, a target organ for PAH-induced carcinogenesis. Past work in this laboratory has concentrated on the metabolism of polycyclic hydrocarbons in susceptible mammalian species, such as mouse and rabbit, and on tissues which are sensitive to chemical carcinogenesis by PAH, such as skin. Recently these studies were expanded to investigate the stereoselectivity of PAH activation in these systems using the technique of CSP HPLC, and reports have been published on this aspect of chrysene metabolism in the skin of rodents and humans[46] and of BP metabolism by different tissues of the mouse[47]. Stereoselectivity of the formation of BP dihydrodiols by isolated cytochromes P-450 from rat liver has also been investigated[48]. In most cases the (+)-enantiomers of dihydrodiols of PAH are less strongly retained by Pirkle column packings and elute first. However, it has been noted that the elution order of the enantiomers of *trans*-9,10-dihydro-9,10-dihydroxybenzo(a)pyrene (BP-9,10-dihydrodiol) on the ionically- and covalently-bonded phenylglycine-based columns is such that the (+)-9*S*,10*S* enantiomer is more strongly retained, observations based on circular dichroism data and on measurements of optical rotation of the resolved enantiomers ([α_D]MeOH = -168° and +188° for the first- and second-eluting peaks, respectively). This elution order is the reverse of that found for BP-7, 8-dihydrodiol, and is also in contradiction with previous reports [27,29]. It has recently transpired that the order of elution of BP-9,10-dihydrodiol enantiomers was incorrectly assigned in these earlier publications (S.K. Yang, personal communication). The following account describes, as examples of the application of CSP HPLC, similar analyses of the stereoselectivity of BP dihydrodiol formation by samples of human skin and of BP and anthracene dihydrodiol formation by microsomes prepared from the liver and skin of the rabbit.

As mentioned above, human skin is believed to be a target organ for PAH-induced carcinogenesis[2]. Since the stereoselectivity of BP metabolism may be important in determining its carcinogenic potential this was investigated in samples of normal skin from eleven individuals. The occurrence of interindividual variations in this process was also considered.

Rabbit skin has also long been known to be a target tissue[49]. An early study[50] showed that rabbits fed anthracene in their diet excreted *trans*-1,2-dihydro-1,2-

dihydroxyanthracene (anthracene-1,2-dihydrodiol) in their urine as the (+)-enantiomer, whereas rats maintained on a similar diet produced the (-)-enantiomer. More recently Von Tungeln and Fu[35] extracted the 1,2-dihydrodiol largely as the (-)-1R,2R enantiomer following incubations of anthracene with rat liver microsomes. In view of these observations the work described below was conducted in order to determine the stereoselectivity of anthracene and BP metabolism by rabbit skin and liver *in vitro*.

EXPERIMENTAL

Chemicals

[G-^{3}H]BP and [*side ring*-U-^{14}C]anthracene were obtained from Amersham International plc, Amersham, Bucks., UK. Unlabelled ($\pm$)-*trans*-BP-4,5,-, -7,8-, and -9,10-dihydrodiols were obtained from NCI Chemical Carcinogen Standard Repository, NIH, Bethesda, MD, USA and purified, unlabelled ($\pm$)-*trans*-anthracene-1,2-dihydrodiol was obtained as described previously[51].

Metabolism of BP by Human Skin in Short-Term Organ Culture

Incubations were conducted essentially as described[52]. Samples of normal human skin from eleven individuals, obtained either from reduction mammoplasty, mastectomy or amputation, were trimmed of excess fat and connective tissue and placed on stainless steel wire grids overlaid with sterile lens paper. The grids were placed into petri dishes containing culture fluid and the skin pieces treated with a topical application of [^{3}H]BP (0.1 μmol/cm^2, 0.33-1.8 mCi/μmol) as a 4 mM solution in acetone. Incubations were carried out at 37°C for 17-20h in an atmosphere of 10% CO_2 in air.

At the end of this time both the skin pieces and culture fluids were removed from the petri dishes and extracted. In the case of skin this was achieved by freezing the samples in liquid nitrogen, scraping the dermal surfaces sufficiently so as to retain only the epidermal layer and then powdering in liquid nitrogen by consecutive use of an MSE homogeniser fitted with a Vortex beaker (100 ml) (MSE Scientific Instruments, Crawley, West Sussex, UK) and a Polytron homogeniser (Kinematica GmbH, Luzern, Switzerland). Powdered tissues were then further homogenised in aqueous 4-aminosalicylic acid (25 ml, 5%, w/v) using a Potter Elvehjem homogeniser. Aqueous sodium dodecyl sulphate (2.5 ml, 10%, w/v) was added and the mixture shaken. Both homogenised skin samples and culture fluids were extracted with ethyl acetate (2 x 1 vol), the extracts dried (Na$_2$SO$_4$), UV-absorbing quantities of non-radioactive BP dihydrodiols added and the mixture evaporated to low volume. The metabolite mixtures were resolved by preparative thin-layer chromatography (TLC) on silica gel plates in benzene/ethanol (9:1, v/v), the bands of silica containing dihydrodiols (Rf $\approx$ 0.4 eluted with ethanol (25 ml) and the extracts filtered prior to HPLC.

Metabolism of BP and Anthracene by Microsomal Fractions of Rabbit Skin and Liver

Microsomal fractions from the liver and aural epidermis of two male New Zealand White rabbits (2 kg) were prepared by differential centrifugation as detailed previously[47]. Incubation of rabbit tissue microsomes with radiolabelled hydrocarbon for 30 min at 37°C was carried out as described[47], the reaction being initiated by addition of an acetone solution (20 μl) of either [^{3}H]BP (4.0 mM, 2.0 mCi/μmol) or [^{14}C]anthracene (4.0 mM, 11.2 mCi/mmol). Following extraction with ethyl acetate (2 x 1 vol), UV-absorbing quantities of non-radioactive BP or anthracene dihydrodiols were added to the metabolite mixtures which were then run on TLC plates as described above. The bands of silica containing dihydrodiols (Rf $\approx$ 0.4 and 0.6 for BP and anthracene, respectively) were eluted with ethanol (25 ml) and the extracts filtered prior to HPLC.

(a) *Dihydrodiols*

Extracts containing mixtures of synthetic, non-radioactive BP or anthracene dihydrodiols and radioactive metabolites were evaporated to dryness under reduced pressure. The residues were redissolved in methanol/water (1:1 and 1:2.3, v/v, for BP and anthracene, respectively) and subjected in their entirety to HPLC using a Waters liquid chromatography system (Waters Associates, Harrow, Middlesex, UK). BP dihydrodiols were analyzed using a Spherisorb 5 ODS column (250 x 4.6 mm) (HPLC Technology Ltd., Macclesfield, Cheshire, UK) eluted at room temperature with a convex gradient (Waters 660 solvent programmer, gradient 3) of 45-60% (v/v) methanol in water over 60 min (1.5 ml/min). Both the anthracene-1,2-dihydrodiol and its 1,2,3,4-tetrahydro derivative were examined using a Zorbax ODS column (250 x 4.6 mm) (HPLC Technology Ltd., Macclesfield, Cheshire, UK) eluted at room temperature with a linear gradient of 30-100% (v/v) methanol in water over 60 min (1.8 ml/min). Eluates were monitored at 254 nm (Waters 440 UV detector), fractions (1 min) of the eluates collected and aliquots (0.1 ml) removed for determination of radioactivity by liquid scintillation counting.

(b) *Stereochemical Analysis of PAH Dihydrodiol Formation*

Direct resolution of the enantiomers of BP-7,8-dihydrodiol obtained from the reversed-phase HPLC separation of metabolic dihydrodiols was performed on a Pirkle 1-A column (250x4.6 mm) (Phase Separations Ltd., Queensferry, Clwyd, UK) with a stationary phase of chiral N-(3,5-dinitrobenzoyl)phenylglycine ionically bonded to γ–aminopropyl silanised silica[47]. A similar separation of the enantiomers of anthracene-1,2-dihydrodiol did not prove possible. However, enantiomers of the *trans*-1,2-dihydroxy-1,2,3,4-tetrahydro-anthracene (anthracene-1,2-tetrahydrodiol), produced by hydrogenation of the dihydrodiol (see below), were resolved on a Prikle 1-A column. The column was eluted with a mixture of hexane/ethanol/acetonitrile in a ratio of either 27:2:1 (by volume) for BP-7,8-dihydrodiol or 37:2:1 (by volume) for anthracene-1,2-tetrahydrodiol (2.0 ml/min). Eluates were monitored at 254 nm, fractions (1 min) collected and radioactivity determined by liquid scintillation counting.

Catalytic Hydrogenation of the Anthracene-1,2-dihydrodiol Metabolite

Catalytic hydrogenation was carried out in a manner similar to that described by Von Tungeln and Fu[35]. Fractions of eluate from the reversed-phase HPLC separation of anthracene-1,2-dihydrodiol identified as containing both synthetic and metabolic dihydrodiol by their UV spectra and radioactive content were pooled and evaporated to dryness. The mixture was taken up in THF/TEA (99:1, v/v, 3 ml), transferred to a 10 ml glass tube and platinum (IV) oxide, type B (5 mg) (Johnson Matthey Chemicals Ltd., Royston, Herts., UK) added. Hydrogen was then bubbled through the solution at atmospheric pressure and ambient temperature for 2h. Following centrifugation the supernatant was removed and the pelleted catalyst washed with acetone (2 x 3 ml). The combined supernatant and washings were evaporated to dryness under reduced pressure and the anthracene-1,2-tetrahydrodiol product purified by reversed-phase HPLC.

RESULTS AND DISCUSSION

Stereoselectivity of BP-7,8-Dihydrodiol Formation by Human Skin in Short-Term Organ Culture

Mixtures of metabolic and non-radioactive BP-7,8-dihydrodiol extracted from the skin and culture fluid of eleven individuals and purified by reversed-phase HPLC as described above were resolved into their constituent enantiomers by CSP HPLC on a Pirkle 1-A column. The degree of resolution in each case was similar to that shown in Fig. 2. Data on the stereochemistry of BP-7,8-dihydrodiol formation from [3H]BP in the skin of these patients are given in Table 1. These show that larger quantities of 7,8-dihydrodiol enantiomers were extracted from the tissue itself than from the culture fluid, possibly because of a propensity of BP-7,8-dihydrodiol for non-covalent binding with subcellular components[53,54]. As a consequence of this the stereoselectivity of the BP-7,8-dihydrodiol extracted from skin and culture fluid combined reflected that of the dihydrodiol extracted from skin alone, with the (-)-7R,8R enantiomer predominating in 9 out of the 10 cases where any 7,8-dihydrodiol was detected. This stereoselectivity of dihydrodiol formation is in agreement with previous studies on PAH activation in skin[46,47]. The exception to this was patient 2 where an approximately racemic mixture was found. In this individual, and also to greater or lesser extents in the other patients, proportionately more (+)- than (-)-enantiomer was extracted from the culture fluid than from the skin itself. This suggests that the 7R,8R form of the 7,8-dihydrodiol may possess a structural conformation more favourable to binding with subcellular components than does the 7S,8S enantiomer.

The differences observed here in the stereoselectivity of BP-7,8-dihydrodiol formation between individuals could be accounted for in several ways. Firstly, microsomal cytochrome P-450 linked mixed function oxidase activity, which has been shown to reside largely within the epidermis of human skin[55,56], is known to be inducible by the application of PAH to skin[57]. Enzyme induction in mouse keratinocytes has been shown to alter the stereoselectivity of the further metabolism of BP-7,8-dihydrodiol[58]. Thus differential exposure to various chemical inducers present in the environment could lead to the induction of distinct enzyme activities and consequently to variations in BP metabolism[59]. Those individuals who, by virtue of their occupation, are believed to be most at risk of developing PAH-induced skin cancer, suffer multiple exposures, often for periods of several years, to mixtures of PAH in the forms of pitches, oils or tars[2]. Secondly, there might exist within the human population, subpopulations that express distinct PAH-metabolising enzymes. Such genetic polymorphism has been described for the oxidation of certain drugs and tenuous links drawn between this and the susceptibility of individuals to PAH-induced carcinogenesis[60,61]. Similarly, certain individuals might be more sensitive to enzyme induction than others[56]. Different areas of skin could conceivably show different selectivities in PAH metabolism, although the limited range of sources of skin used in the present study (Table 1) as well as previous evidence[55] tends to argue against this. Chapman et al,[55] have reported a positive correlation between epidermal microsomal mixed function oxidase activity and patient age, although a more recent study[56] found no correlation of this activity or its inducibility with age, sex, location of domicile or underlying disease of the donor. No attempt was made in the present investigation to correlate the differences observed in the stereoselectivity of BP metabolism with patient history. The contribution of such differences to individual susceptibilities to PAH-induced skin carcinogenesis is at present a matter for conjecture.

Differential Stereoselectivity in the Formation of BP-7,8- and Anthracene-1,2-dihydrodiols by Rabbit Skin and Liver Microsomes

Incubations of [^{3}H]BP and [^{14}C]anthracene with microsomes from rabbit skin and liver were conducted as described above, and metabolic BP-7,8- and anthracene-1,2-dihydrodiols isolated by reversed-phase HPLC. The enantiomers of BP-7,8-dihydrodiol were resolved by HPLC on a Pirkle 1-A column (Fig. 2). Direct resolution of the enantiomers of anthracene-1,2-dihydrodiol did not prove possible (Fig. 3A), but following hydrogenation to its tetrahydro-derivative separation was achieved using a Pirkle 1-A column (Figs. 3B and 4). This phenomenon has been suggested to be due to a more saturated ring being more flexible, hence providing a stronger stereochemical repulsive interaction[29]. Results from the quantification of these enantiomers are presented in Table 2. Both tissue types showed a high degree of metabolic stereoselectivity in the conversion of BP to BP-7,8-dihydrodiol, with the 7R,8R enantiomer predominating in both cases, as reported above for human skin and as found with other species[4,7,47]. In contrast, anthracene-1,2-dihydrodiol extracted from these incubations showed a relatively low optical purity (32% for both liver and skin microsomes) with the 1S,2S enantiomer being the major form detected. Liver microsomes prepared from untreated rats have previously been reported to metabolise anthracene to a 1,2-dihydrodiol derivative having an optical purity of 60% with the 1R,2R enantiomer predominating[35]. The present data therefore confirm the *in vivo* findings of Boyland and Levi[50].

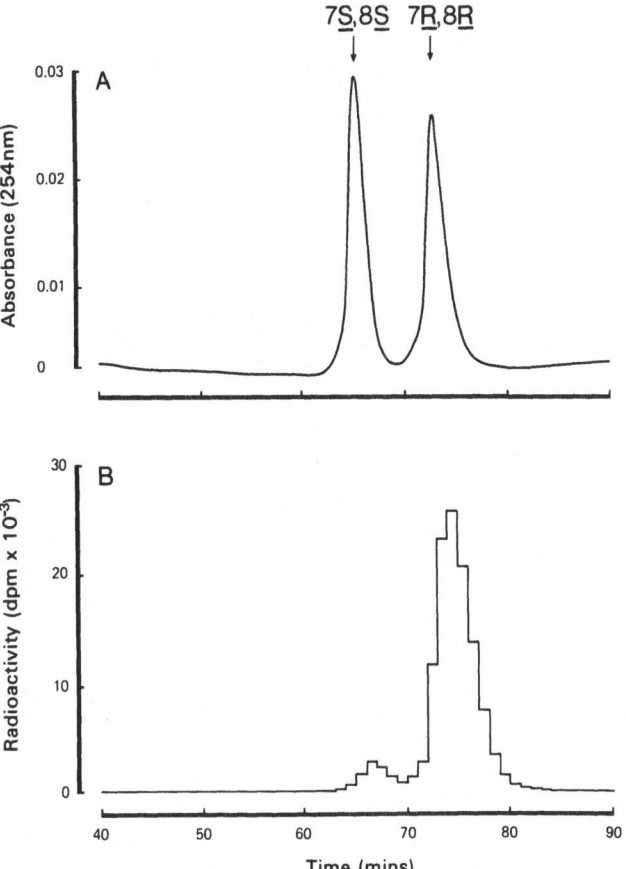

Fig. 2. Resolution by chiral HPLC of the *S,S* and *R,R* enantiomers of BP-7,8-dihydrodiol. A: synthetic, non-radioactive dihydrodiol; B: dihydrodiol extracted from incubation of [³H]BP with rabbit liver microsomes. Pirkle 1-A columns (250 x 4.6 mm) were eluted isocratically with a mixture of hexane/ethanol/acetonitrile (27:2:1, by volume) as described in the text and the eluates were examined for the presence of UV-absorbing materials and radioactivity. The enantiomers were assigned as *S,S* or *R,R* according to Yang *et al*,[27].

Differences observed in the stereoselective formation of PAH dihydrodiols in microsomal incubations between differently-induced animals or different species must be due to variations in the selectivities of the enzymes present. Studies performed with isolated rat liver isozymes of cytochrome P-450 have shown that the major 3-MC-inducible form, P-450c[62], metabolised anthracene to its 1,2-dihydrodiol predominantly (99%) as the 1*R*,2*R* enantiomer. This was irrespective of the amounts of epoxide hydrolase present[17,18]. The stereoselectivity of anthracene-1,2-dihydrodiol formation by cytochrome P-450b, a phenobarbital-inducible form[62], was very much dependent, however, on the amount of epoxide hydrolase added to the system, such that at high epoxide hydrolase concentrations the 1*S*,2*S* enantiomer was the major form of the dihydrodiol extracted[18]. Although there is evidence for structural inter-relatedness of some isozymes of cytochrome P-450 detected in livers of rabbit and rat[63], this may not be true for all forms. Hence the contrasting stereoselectivity of anthracene-1,2-dihydrodiol formation reported here compared with that observed previously for rat liver microsomes could be caused by interspecies differences in the cytochrome P-450 profile and/or the relative amounts of P-450 and epoxide hydrolase

Table 1. Stereoselectivity of BP-7,8-Dihydrodiol Extracted From Culture Fluids and Tissue Samples Following Topical Application of [3H]BP to Human Skin *in vitro*

Patient No.	Sex	Source of Tissue*	From culture fluid			BP-7,8-dihydrodiol enantiomers extracted (fmol/cm² skin) From skin			Overall ratio [%(+):(-)]
			(+)	(-)	[%(+):(-)]	(+)	(-)	[%(+):(-)]	
1	Female	R	107.4	96.6	53:47	2060.0	10250.0	17:83	17:83
2	Male	A	624.6	27.7	96:4	2020.0	2050.0	50:50	56:44
3	Male	A	4.0	78.0	5:95	5.8	1840.0	<1:>99	1:99
4	Female	M	2.4	4.9	33:67	7.0	1865.0	<1:>99	<1:>99
5	Female	A	9.1	162.5	5:95	37.8	15127.0	<1:>99	<1:>99
6	Female	M	43.3	4.6	90:10	10.8	2050.0	<1:>99	3:97
7	Female	M	2.3	44.9	5:95	25.5	970.8	3:97	3:97
8	Female	M	10.7	118.9	8:92	54.0	3960.0	1:99	2:98
9	Female	M	7.0	5.9	54:46	20.0	1640.0	1:99	2:98
10	Female	A	6.5	56.2	10:90	100.0	2980.0	3:97	3:97
11	Female	A	-**	-	N.D.***	-	-	N.D.	N.D.

* R, reduction mammoplasty; M, mastectomy; A, amputation. ** -, not detected. *** N.D., not determined.

Solvent: 7.5% {EtOH:MeCN, 2:1}
in hexane

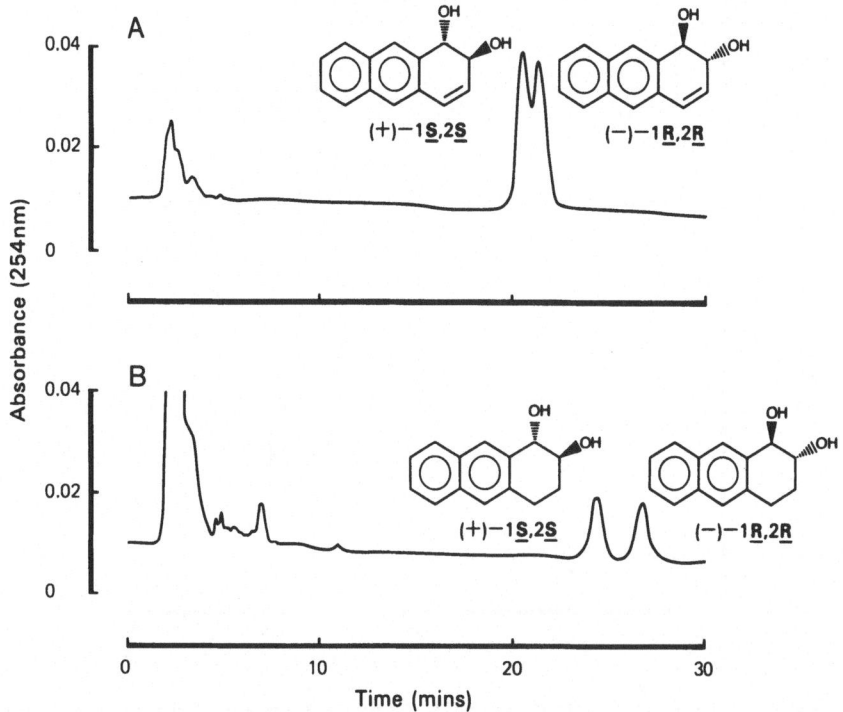

Fig. 3.   Separation by chiral HPLC on a Pirkle 1-A column (250 x 4.6 mm) of the *S,S* and *R,R*
enantiomers of A: anthracene-1,2-dihydrodiol; B: anthracene-1,2-tetrahydrodiol.
Columns were eluted isocratically with a mixture of hexane/ethanol/acetonitrile
(37:2:1, by volume) as described in the text. Elution order of the enantiomers was
assigned according to Von Tungeln and Fu[35].

activity present in the microsomes. These possibilities may be worthy of further investigation.

The higher stereoselectivity of BP-7,8-dihydrodiol formation compared with
anthracene-1,2-dihydrodiol formation noted here might be due either to a 'tighter fit' of the
larger BP molecule in the cytochrome P-450 binding site or to distinct activities being
involved in their metabolism. Again, such possibilities could be resolved by studies on the
isolated enzymes concerned.

CONCLUSIONS

The methodologies which have been employed in the HPLC separation of enantiomers
of epoxide and dihydrodiol metabolites of various PAH, and the employment of these
techniques have been briefly reviewed. The development of Pirkle chiral stationary phases
has proved particularly useful for this purpose as it allows direct and rapid resolution of
enantiomers without the need for prior derivatization, a factor of prime importance when
only small amounts of metabolite are available for analysis. Two examples of the use of
Pirkle columns in the elucidation of the stereoselective metabolism of PAH have been given
from recent work conducted in these laboratories, viz activation of BP by human skin *in
vitro* and of BP and anthracene by microsomes prepared from skin and liver of the rabbit. In
both cases radioactively-labelled substrates were employed and the resulting radioactively-
labelled dihydrodiol metabolites resolved and quantified using a Pirkle 1-A column, further
demonstrating the sensitivity of this technique. It also extends its usefulness beyond

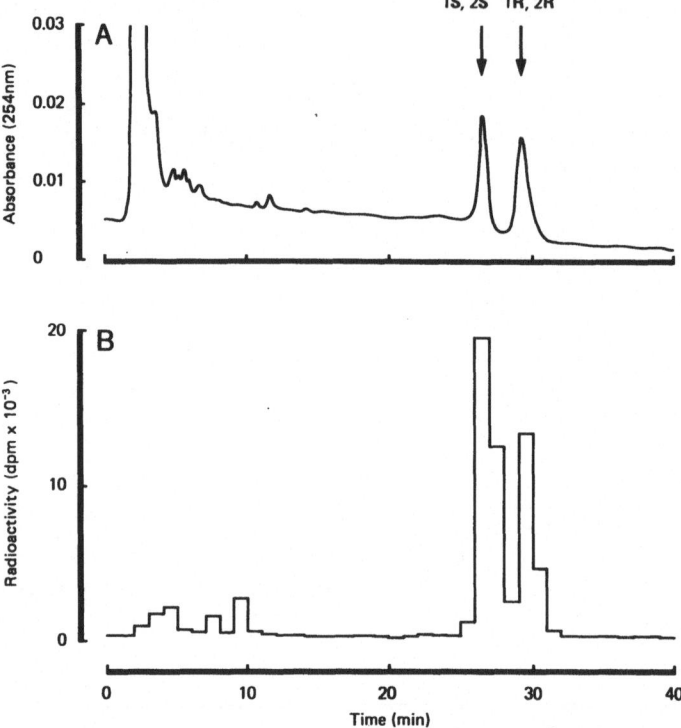

Fig. 4. Resolution by chiral HPLC of the *S,S* and *R,R* enantiomers of anthracene-1,2-tetrahydrodiol. A: synthetic, non-radioactive tetrahydrodiol; B: tetrahydrodiol resulting from hydrogenation of dihydrodiol extracted from incubation of [^{14}C] anthracene with rabbit liver microsomes. HPLC conditions as for Fig. 3.

Table 2. Stereoselectivity of BP-7,8- and Anthracene-1,2-Dihydrodiols Extracted Following Incubation of [^{3}H]BP and [^{14}C]Anthracene with Rabbit Liver and Skin Microsomal Fractions

| Dihydrodiol | Tissue | Amount of Metabolic Dihydrodiol Enantiomers Extracted (fmol/min/mg microsomal protein) | | |
		(+)-*S,S*	(−)-*R,R*	Ratio [%(+):(−)]
BP-7,8-Dihydrodiol	Liver	10.1	182.3	5:95
	Skin	0.1	2.6	4:96
Anthracene-1,2-Dihydrodiol	Liver	11,467.0	5,796.0	66:34
	Skin	138.6	71.2	66:34

the range of applications developed by Yang et al, who have thus far resolved only non-radioactive metabolites on these systems.

Acknowledgements

The authors wish to thank Dr. S.K. Yang for helpful discussions, Professors G. Westbury and B. Gusterson and Dr. R. L. Carter of the Royal Marsden Hospital, London for supplying human skin samples, and Mrs. D. K. Parker for valuable technical assistance. This work was supported in part by PHS grant CA 21959 awarded by the National Cancer Institute and, in part, by grants to the Institute of Cancer Research from the Medical Research Council and the Cancer Research Campaign.

REFERENCES

1. IARC Monographs on the Evaluation of the Carcinogenic Risk of Chemicals to Man, Vol. 3: Certain polycyclic aromatic hydrocarbons and heterocyclic compounds, International Agency for Research on Cancer, Lyon (1973).
2. S. A. Henry, *Br.Med.Bull.*, 4:389 (1947).
3. D. H. Phillips and P. L. Grover, in: "Monitoring Human Exposure to Carcinogenic and Mutagenic Agents," A. Berlin, M. Draper, K. Hemminki and H. Vainio, eds., International Agency for Research on Cancer, Lyon, p. 47 (1984).
4. D. R. Thakker, H. Yagi, W. Levin, A. W. Wood, A. H. Conney and D. M. Jerina, in: "Bioactivation of Foreign Compounds," M. W. Anders, ed., Academic Press, Inc., New York, ch.7, p. 177 (1985).
5. P. Sims, P. L. Grover, A. Swaisland, K. Pal and A. Hewer, *Nature*, 252:326 (1974).
6. A. H. Conney, *Cancer Res.*, 42:4875 (1982).
7. S. K. Yang, M. Mushtaq and P.-L. Chiu, in: "ACS Symposium Series No. 283, Polycyclic Hydrocarbons and Carcinogenesis," R.G. Harvey, ed., American Chemical Society, Washington D.C., p. 19 (1985).
8. P. J. van Bladeren, R. N. Armstrong, D. Cobb, D. R. Thakker, D. E. Ryan, P. E. Thomas, N. D. Sharma, D. R. Boyd, W. Levin and D. M. Jerina, *Biochem. Biophys. Res. Commun.*, 106:602 (1982).
9. S. K. Yang, P. P. Roller and H. V. Gelboin, *Biochemistry*, 16:3680 (1977).
10. D. R. Thakker, H. Yagi, W. Levin, A. Y. H. Lu, A. H. Conney and D. M., Jerina, *J.Biol.Chem.*, 252:6328 (1977).
11. H. Yagi and D. M. Jerina, *J.Am.Chem.Soc.* 104:4026 (1982).
12. D. R. Thakker, W. Levin, H. Yagi, M. Tada, D. E. Ryan, P. E. Thomas, A. H. Conney and D. M. Jerina, *J.Biol.Chem.*, 257:5103 (1982).
13. M. W. Chou, P.-L. Chiu, P. P. Fu and S. K. Yang, *Carcinogenesis*, 4:629 (1983).
14. H. Yagi, D. R. Thakker, Y. Ittah, M. Croisy-Delcey and D. M. Jerina, *Tetrahedron Lett.*, 24:1349 (1983).
15. Y. Ittah, D. R. Thakker, W. Levin, M. Croisy-Delcey, D. E. Ryan, P. E. Thomas, A. H. Conney and D. M. Jerina, *Chem.-Biol.Interact.*, 45:15 (1983).
16. P. J. van Bladeren, S. K. Balani, J. M. Sayer, D. R. Thakker, D. R. Boyd, D. E. Ryan, P. E. Thomas, W. Levin and D. M. Jerina, *Biochem.Biophys.Res.Commun.*, 145:160 (1987).
17. P. J. van Bladeren, K. P. Vyas, J. M. Sayer, D. E. Ryan, P. E. Thomas, W. Levin and D. M. Jerina, *J.Biol.Chem.*, 259:8966 (1984).
18. P. J. van Bladeren, J. M. Sayer, D. E. Ryan, P. E. Thomas, W. Levin and D. M. Jerina, *J.Biol.Chem.*, 260:10226 (1985).
19. W. Levin, M. K. Buening, A. W. Wood, R. L. Chang, B. Kedzierski, D. R. Thakker, D. R. Boyd, G. S. Gadaginamath, R. N. Armstrong, H. Yagi, J. M. Karle, T. J. Slaga, D. M. Jerina and A. H. Conney, *J.Biol.Chem.*, 255:9067 (1980).
20. R. N. Armstrong, B. Kedzierski, W. Levin and D. M. Jerina, *J.Biol.Chem.*, 256:4726 (1981).
21. D. M. Jerina, J. M. Sayer, H. Yagi, P. J. van Bladeren, D. R. Thakker, W. Levin, R. L. Chang, A. W. Wood and A. H. Conney, in: "Microsomes and Drug Oxidations," A.R.

Boobis, J. Caldwell, F. De Matteis and C.R. Elcombe eds., Taylor and Francis, London, p.310 (1985).

22. I. W. Wainer and T. D. Doyle, *Liq.Chromatogr.*, 2 (1984).
23. H. B. Weems and S. K. Yang, *Anal.Biochem.*, 125:156 (1982).
24. W. H. Pirkle, D. W. House and J. M. Finn, *J.Chromatogr.*, 192:143 (1980).
25. W. H. Pirkle and J. M. Finn, *J.Org.Chem.*, 46:2935 (1981).
26. S. K. Yang and X.-C. Li, *J.Chromatogr.*, 291:265 (1984).
27. S. K. Yang, H. B. Weems, M. Mushtaq and P. P. Fu, *J.Chromatogr.*, 316:569 (1984).
28. H. B. Weems, M. Mushtaq and S. K. Yang, *Anal.Biochem.*, 148:328 (1985).
29. S. K. Yang, M. Mushtaq, H. B. Weems and P. P. Fu, *J.Liq.Chromatogr.*, 9:473 (1986).
30. P.-L. Chiu and S. K. Yang, *J.Liq.Chromatogr.*, 9:701 (1986).
31. S. K. Yang, M. Mushtaq and P. P. Fu, *J.Chromatogr.*, 371:195 (1986).
32. H. B. Weems, M. Mushtaq, P. P. Fu and S. K. Yang, *J.Chromatogr.*, 371:211 (1986).
33. M. Mushtaq and S. K. Yang, *Carcinogenesis*, 8:705 (1987).
34. S. K. Yang, M. Mushtaq and H. B. Weems, *Arch.Biochem.Biophys.*, 255:48 (1987).
35. L. S. Von Tungeln and P. P. Fu, *Carcinogenesis*, 7:1135 (1986).
36. H. B. Weems, P. P. Fu and S. K. Yang, *Carcinogenesis*, 7:1221 (1986).
37. S. K. Yang and P. P. Fu, *Chem.-Biol.Interact.*, 49:71 (1984).
38. S. K. Yang, M. Mushtaq, H. B. Weems and P. P. Fu, *Tetrahedron Lett.*, 27:433 (1986).
39. S. K. Yang, M. Mushtaq, H. B. Weems, D. W. Miller and P. P. Fu, *Biochem.J.*, 245:191 (1987).
40. S. K. Yang and H. B. Weems, *Anal.Chem.*, 2658 (1984).
41. S. K. Yang and P. P. Fu, *Biochem.J.*, 223:775 (1984).
42. M. Mushtaq, H. B. Weems and S. K. Yang, *Biochem.Biophys.Res.Commun.*, 125:539 (1984).
43. S. K. Balani, H. J. C. Yeh, D. E. Ryan, P. E. Thomas, W. Levin and D. M. Jerina, *Biochem.Biophys.Res.Commun.*, 130:610 (1985).
44. P. P. Fu and S. K. Yang, *Carcinogenesis*, 4:979 (1983).
45. P. P. Fu, L. S. Von Tungeln and M. W. Chou, *Mol.Pharmacol.*, 28:62 (1985).
46. A. Weston, R. M. Hodgson, A. J. Hewer, R. Kuroda and P. L. Grover, *Chem.-Biol.Interact.*, 54:223 (1985).
47. M. Hall and P. L. Grover, *Chem.-Biol.Interact.*, 59:265 (1986).
48. M. Hall, D. K. Parker, M. Christou, C. R. Jefcoate and P. L. Grover *in*: "Drug Metabolism - From Molecules to Man," D. Benford, G.G. Gibson and J.W. Bridges, eds., Taylor and Francis, London, p.411 (1987).
49. K. Yamagiwa and K. Ichikawa, *Verh.Jap.Pathol.Ges.*, 5:142 (1915).
50. E. Boyland and A. A. Levi, *Biochem.J.*, 29:2679 (1935).
51. P. Sims, *Biochem.J.*, 92:621 (1964).
52. A. Weston, P. L. Grover and P. Sims, *Chem.-Biol.Interact.*, 45:359 (1983).
53. R. Mehta, M. Meredith-Brown and G. M. Cohen, *Chem.-Biol.Interact.*, 28:345 (1979).
54. A. Weston, P. L. Grover and P. Sims, *Chem.-Biol.Interact.*, 42:233 (1982).
55. P. H. Chapman, M. D. Rawlins and S. Shuster, *Br.J.Clin.Pharmacol.*, 7:499 (1979).
56. T. Kuroki, J. Hosomi, K. Chida, J. Hosoi and N. Nemoto, *Gann*, 78:45 (1987).
57. W. Levin, A. H. Conney, A. P. Alvares, I. Merkatz and A. Kappas, *Science*, 176:419 (1972).
58. T. Eling, J. Curtis, J. Battista and L. J. Marnett, *Carcinogenesis*, 7:1957 (1986).
59. G. Ekström, C. von Bahr, H. Glaumann and M. Ingelman-Sundberg, *Acta Pharmacol. et Toxicol.*, 50:251 (1982).
60. R. Ayesh, J. R. Idle, J. C. Ritchie, M. J. Crothers and M. R. Hetzel, *Nature*, 312:169 (1984).
61. F. P. Guengerich, M. V. Martin, P. H. Beaune, P. Kremers, T. Wolff and D. J. Waxman, *J.Biol.Chem.*, 261:5051 (1986).
62. D. E. Ryan, P. E. Thomas, L. M. Reik and W. Levin, *Xenobiotica*, 12:727 (1982).
63. D. Sesardic, A. R. Boobis, J. McQuade, S. Baker, E. A. Lock, C. R. Elcombe, R. T. Robson, C. Hayward and D. S. Davies, *Biochem.J.*, 236:569 (1986).

A NOTE ON THE ROLE OF SOLVENTS AND STERIC FACTORS IN THE

RESOLUTION OF β–BLOCKER DRUGS ON CHIRAL UREA PHASES

Nagaraja K. R. Rao, Robert C. Towill and Bindu Todd

Jones Chromatography Ltd
Tir-y-Berth Industrial Estate, New Road
Hengoed, Mid Glamorgan CF8 8AU, UK

INTRODUCTION

A wide range of β-aminoalcohols of the type Ar-O-CH2-CH(OH)-CH2-CH-NH(CH3)2 (where Ar is an aromatic group) possess pharmacological activity and a number of them are used as β-blocking agents for the treatment of hypertension. These drugs are commonly administered as a racemic mixture even though it has been known for sometime[1] that the L form shows a 50 to 500 fold greater activity than the D form. Furthermore the work of Nelson and Burke[2] demonstrated that some of the metabolites of the D form showed signs of toxicity. In view of these findings, there has been a natural interest in the resolution of these drugs and their metabolities.

There are several HPLC methods reported in the literature for the resolution of these - blocker drugs[3-6] involving the use of highly toxic phosgene for derivatisation. Recently, Propranolol and Oxyprenolol were reported[7] to be resolved without derivatisation although it appears that this method is only applicable to these two compounds.

In the course of our study[8] concerning the preparation of novel chiral stationary phases, we observed that a number of the β-blocker drugs could be resolved on chiral urea stationary phases and the resolution largely depended on the nature of the substituents on the chiral as well as the urea centers. Here we describe details of our findings and present a rationale for the resolution mechanism.

EXPERIMENTAL

All the chiral amino acid derivatives listed in the Table 1 were prepared according to established procedures[9] and the new naphthyl derivatives were fully characterized by analytical, spectroscopic and optical methods. Chiral stationary phases were prepared by linking the terminal carboxyl group to an amino function on Apex 5μ silica through activation of the carboxyl group. The chiral phases were packed into 250 x 4.5mm I.D. stainless tubes by the methods specially developed by Jones Chromatography Ltd. (Hengoed, Mid Glamorgan, UK). The chromatography was performed using an Altex 100 A pump, a Rheodyne 7125 valve injector and the effluents were monitored at 254nm.

All the solvents used in this study were of analytical grade and the isocyanates were obtained from Aldrich Chemical Co. t-Butyl, phenyl and naphthyl derivatives of Propranolol (D,L and the racemic DL) were prepared by heating a solution of the drug in anhydrous tetrahydrofuran in a vial sealed under nitrogen at 80°C with 2.1 equivalent of the appropriate isocyanate (Aldrich Chemical Co.) for two hours. After cooling, the excess reagent was destroyed by treating with methanol. Removal of the solvent afforded a partially crystalline solid and a solution of this in chloroform was used directly in the analysis. Field desorption mass spectrometric data of these derivatives indicated that both hydroxyl and amino protons had reacted.

RESULTS AND DISCUSSIONS

Twelve chiral urea stationary phases (CSPs) having a general structural formula R-CH-NH-CO-NHR' were considered suitable for this study. The main criteria of selection were, firstly the materials could be prepared in large quantities and in high enantiomeric purity, and secondly, the nature of the substituents on the chiral center as well as the urea functionality could be easily manipulated. Thus, the groups R and R' (on the CSP) could be aromatic in nature (R and R' = phenyl or naphthyl; CSP5, 6, 8 and 9) or aliphatic and capable of donating elections freely (R and R' = isopropyl or t-butyl CSP1 and 10), or a combination of these (CSP2, 3, 4, 7, 11 and 12). The 3,5-dinitrobenzoyl leucine phase was also included in this study with a view to observing the effects of an aromatic amide functional group on resolution. (CSP 13).

In the table a summary of the chromatographic behavior of derivatised propranolol on these phases is shown. Of the three different types of derivatives studied, only the t-butyl analogue was resolvable. The resolution depended entirely on the nature of groups R and R'. With the change of aliphatic character at the chiral center of the CSP (group R) to aromatic the magnitude of separation dramatically dropped. The opposite effect was observed when changes were made at the urea terminal. Thus the results clearly pointed to the fact that for the best resolution an aliphatic group was needed at the chiral center on the amino acid with an aromatic urea derivative. The separation of the t-butyl propranolol derivative on different stationary phases is shown in Fig. 1.

Several interactions such as dipolar stacking, hydrogen bonding[9,10] etc., have been proposed as major contributing factors in the resolution process when a urea moiety is involved. Other interactions such as π-π were considered[11] less significant. In the present case, the nature of various substituents appears to be very critical for any separations to occur and this demonstrated that hydrogen bonding interactions were in fact prominent. Two different orientations (represented in Fig. 2) of the analyte molecule with respect to the

Table 1.   Resolution of Bis-t-butyl Derivatives of Propranalol on Various Stationary Phases. Mobile phase 1.5% isopropanol in hexane flow rate 1 ml/min. Under these conditions the aromatic derivatives did not resolve.

Stationary Phase		Resolution Factor
CSP1.	Leucine-t-butyl urea	1.03
CSP2.	Leucine-naphthyl urea	1.20
CSP3.	Leucine-phenyl urea	1.08
CSP4.	Phenylalanine-t-butyl urea	1.00
CSP5.	Phenylalanine-naphthyl urea	1.02
CSP6.	Phenylalanine-phenyl urea	1.00
CSP7.	Phenylalanine-t-butyl urea	1.02
CSP8.	Phenylglycine-naphthyl urea	1.00
CSP9.	Phenylglycine-phenyl urea	1.00
CSP10.	Valine-t-butyl urea	1.02
CSP11.	Valine-napthyl urea	1.07
CSP12.	Valine-phenyl urea	1.15
CSP13.	Valine-3,5-dinitrobenzoyl	1.00

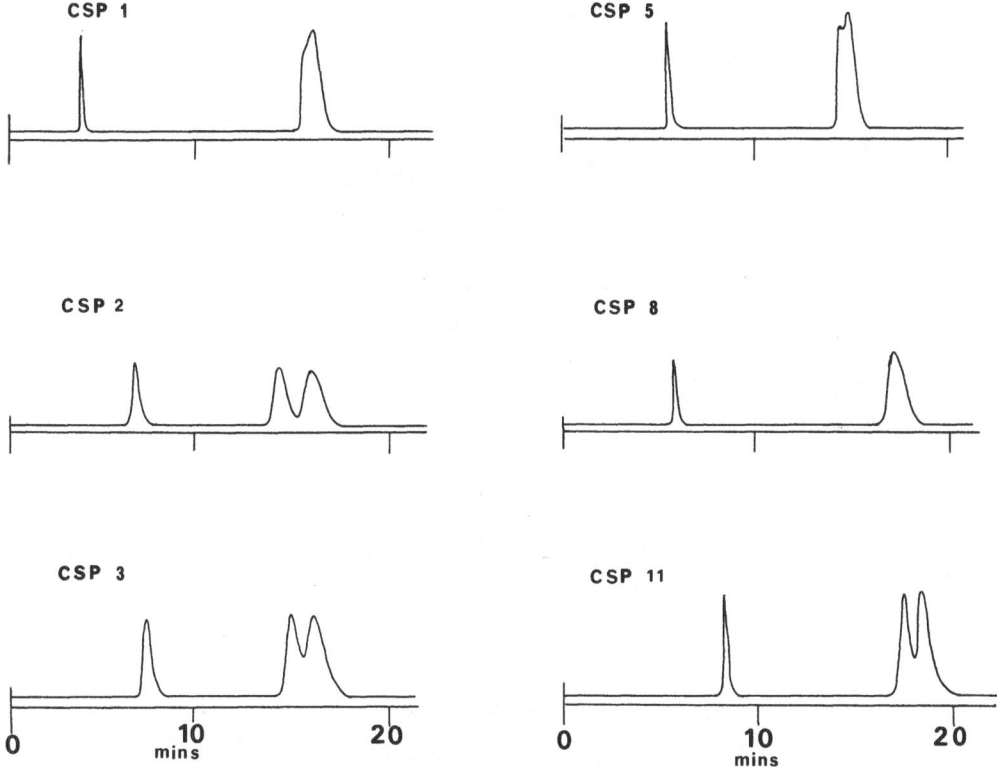

Fig. 1. Resolution of t-butyl derivitives of propranolol on six chiral stationary phases under standard conditions. The L-enantiomer was the least retained in all these cases.

chiral phase may be considered. When the overlap of π aromatic groups are greatest (Fig. 2A), the alkyl group at the chiral center (on the CSP) experiences severe non-bonded interaction with the bulky t-butyl group and this destabilizes the associations required for resolution. The 3,5-dinitrobenzoyl leucine phase where π–π interactions are stronger failed to resolve the analyte. In orientations such as 2B on the other hand, these non-bonded interactions becomes less severe. Molecular model and molecular graphic studies also suggest that the maximum overlap of interactive groups are possible only in models such as 2B and the chiral molecule is able to resolve the racemic analyte.

The applicability of the above described method was extended to other β blocker drugs. Typical separations obtained for the drugs; Alprenolol, Oxyprenolol and Propranolol are given in Fig. 3.

We also explored the possibility of changing solvent selectivity using the selectivity triangle[12]. However, we were unable to obtain any satisfactory separations in a variety of solvent mixtures, ie. hexane/chloroform, hexane/ethyl acetate, hexane/ether.

In conclusion, the chiral urea stationary phase has been shown to be very effective in the separation of β-blocker drugs. The elution order is consistent throughout and the resolution method may be extended to related compounds. The derivatisation method is simple and can be applied to biological samples.

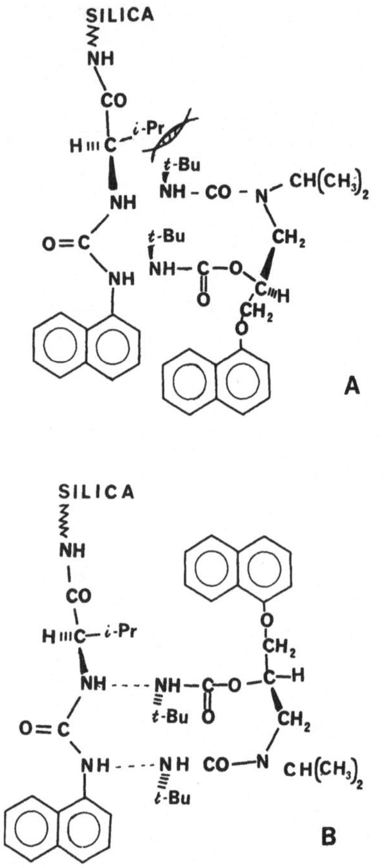

SILICA

A

H⫶⫶C—*i*-Pr
NH
O=C
NH

t-Bu
NH–CO–N⟨CH(CH₃)₂
                CH₂
t-Bu
NH–C–O–C⫶⫶H
      O   CH₂
          O

SILICA

H⫶⫶C—*i*-Pr
NH----NH–C–O–C–H
        t-Bu O   CH₂
O=C
NH----NH CO—N CH(CH₃)₂
   t-Bu

B

Fig. 2.   Schematic representation of two chiral recognition processes.

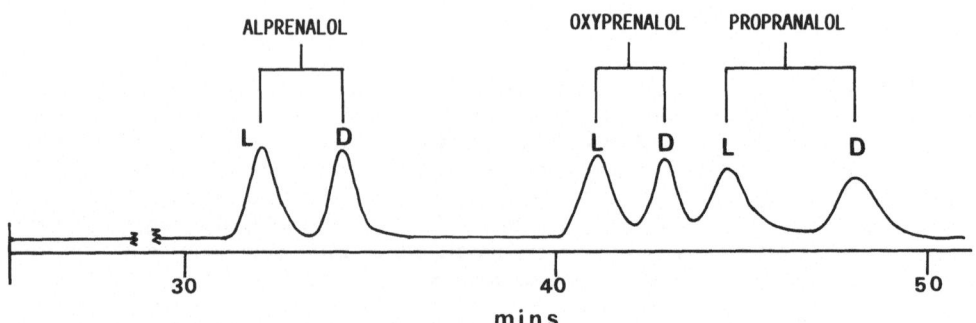

ALPRENALOL   OXYPRENALOL   PROPRANALOL

L   D   L   D   L   D

30   40   5 0

mins

Fig. 3.   Simultaneous resolution of three B-blocker drugs on CSP 12 under the conditions described earlier except that the flow rate was reduced to 0.5 ml/min. The D isomer was more retained in all these cases.

Acknowledgement

    We thank Dr. K.R. Brain of University of Wales Science and Technology, Cardiff for providing samples. Dr. P. A. Williams and Professor D. E. Games of the University College, Cardiff, for circular Dichroic and Mass Spectral determinations.

REFERENCES

1. A. M. Barrett and V. A. Cullum, *J.Pharmacol.*, 34:43 (1968).
2. W. L. Nelson and T. R. Burke Jr., *J.Org.Chem.*, 43:3641 and references cited therein (1978).
3. W. H. Pirkle, J. M. Finn, J. L. Schreiner and B. C. Hamper, *J.Am.Chem.Soc.*, 103:2055 (1984).
4. I. W. Wainer, T. D. Doyle, K. H. Donn and J. R. Powell, *J.Chromatogr.*, 306:405 (1984).
5. J. Hermansson, *J.Chromatogr.*, 325:379 (1985).
6. J. Schulze and W. A. Konig, *J.Chromatogr.*, 355:165 (1986).
7. G. Schill, I. W. Wainer and S. A. Barkan, *J.Liq.Chromatogr.*, 9:641 (1986).
8. Y. Okamoto, M. Kawashima, R. A. Buratami H. Hatada, T. Nashiyama and M. Masuda, *Chem.Letters*, 1237 (1986).
9. N. Oi and Kitahara, *J.Chromatogr*, 285:198 (1985).
10. W. H. Pirkle and M. H. Hyun, *J.Chromatogr.*, 322:295 (1985).
11. N. Oi, H. Kitahara and R. Osumi, *Nippon Kagaku Karishi,* 7:999 (1986).
12. L. R. Snyder and J. J. Kirkland, Introduction to Modern Liquid Chromatography, 2nd Edition, Wiley, New York, pp. 260-65 (1979).

A NOTE ON COMPLEXATION OF DANSYL AMINO ACID ENANTIOMERS

BY β-CYCLODEXTRIN STUDIED BY HIGH PERFORMANCE LIQUID

CHROMATOGRAPHY AND FLUORESCENCE MEASUREMENTS

David A. Briggs, Roger B. Homer and Russell Godfrey*

School of Chemical Sciences, University of East Anglia,
Norwich NR4 7 TJ, UK and *Cyanamid of Great Britain Ltd.
Fareham Road, Gosport, Hampshire PO13 0AS, UK

It has been shown recently[1] that dansyl (1-dimethylaminonaphthalene-5-sulphonyl) amino acids can be resolved on a reversed-phase HPLC column by incorporation of β-cyclodextrin into the mobile phase. We have sought to elucidate the structural and energetic factors involved through a chromatographic determination of the association constants for the D- and L- dansyl amino acid/cyclodextrin complexes, combined with a parallel fluorescence study of the complexation.

The reduction in retention time, on a 10 cm C8 column, and increase in resolution as the cyclodextrin concentration in the mobile phase (pH 7.0 phosphate buffer 0.01M, 10% acetonitrile) is increased, is shown in Fig. 1. In all cases the D-enantiomer elutes first.

The retention time, t_{obs} depends on the retention times of the free dansyl amino acid, t_0 and its cyclodextrin complex, t_{cyd}, and the association for complex formation, K. It is related to the concentration of cyclodextrin,[CD] in the mobile phase by the following equation[2],

$$t_{obs} = (t_0 + t_{cyd} K[CD])/(1 + K[CD]) \qquad (1)$$

If the cyclodextrin complex is not retained on the column, $t_{cyd} \approx 0$, then this equation can be reduced to

$$t_0/t_{obs} = 1 + K[CD] \qquad (2)$$

The data are plotted in this form in Fig. 2. The linearity of the lines indicates that the predominant species is the 1:1 complex, and the intercept of unity confirms that the complex is not retained on the column. The association constants are gathered in the Table. They do not differ greatly but in each case the D enantiomer forms the more stable complex.

The fluorescence emission spectra of dansyl derivatives are sensitive to solvent polarity, with the intensity increasing as the polarity decreases[3]. Complexation of the dansyl amino acids by β–cyclodextrin in aqueous solution produces a substantial increase in intensity[4] which we attribute to the inclusion of the fluorophore into the non-polar cavity of the cyclodextrin. This accounts for the similarity of the association constants, the interaction of the amino acid residue with the chiral sites on the rim of the cyclodextrin then provides the source of the chiral discrimination seen in the Table.

The increase in fluorescence intensity on complexation provides an independent method of measuring the association constant, but the presence of organic solvent which is

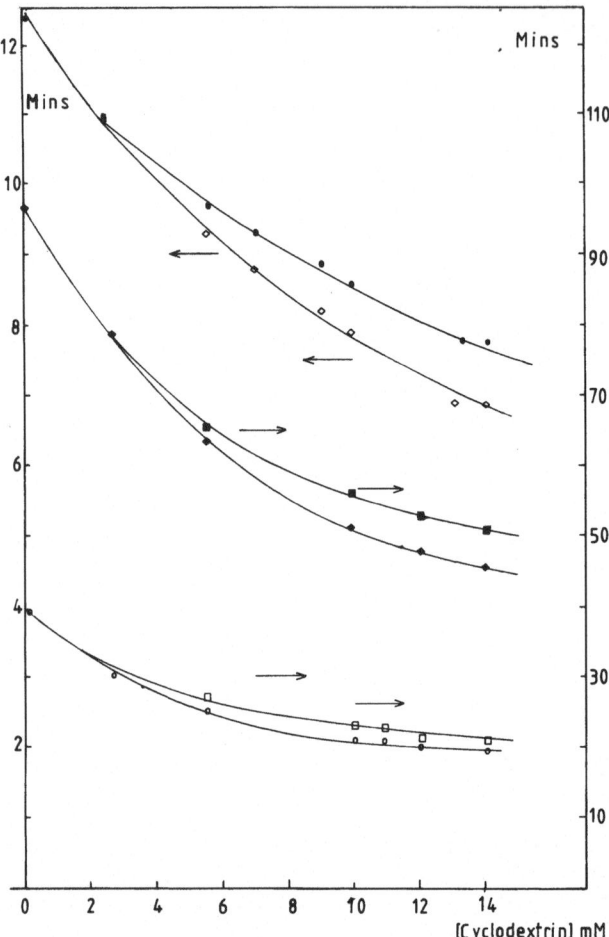

Fig. 1. The effect of increasing β-cyclodextrin concentration on the retention time and resolution of some dansyl-D, L-aminoacids using a 100 x 4.6 mm 5µ spherisorb Octyl column with 10% acetonitrile, 90% pH 7.0 phosphate buffer (0.01M) at 2.0 ml/min, 22 ± 2°C, detection at 215 nm.

● DNS-L-Thr, ◆ DNS-D-Thr, ■ DNS-L-Leu, ◇ DNS-D-Leu, □ DNS-L-Val,
○ DNS-D-Val.

necessary to give good chromatography, limits the increase (Fig. 3). There are two reasons for this, the organic solvent reduces the polarity of the external solvent towards that of the cyclodextrin cavity, thus increasing the fluorescence intensity, and the solvent competes with the fluorophore to bind in the cavity[5], thus reducing the concentration of fluorophore bound. At the acetonitrile concentration used for the chromatography (10%) it is still possible to measure the association constants reasonably precisely.

Where F_0 and F are the experimental fluorescence intensities in the absence and presence of cyclodextrin, and F_{cyd} is the relative fluorescence intensity of the complex. A plot of the equation,

$$1/(F_0 - F) = 1/K[CD](F_0 - F_{cyd}) + 1/(F - F_{cyd}) \qquad (3)$$

allows the association constant (K) and the relative fluorescence intensity of the complex (F_{cyd}) to be calculated from he slope and intercept. The association constants for the dansyl-L-amino acids (and glycine) are collected in the Table, where reasonable agreement with the chromatographic data confirms that both techniques are measuring the same process.

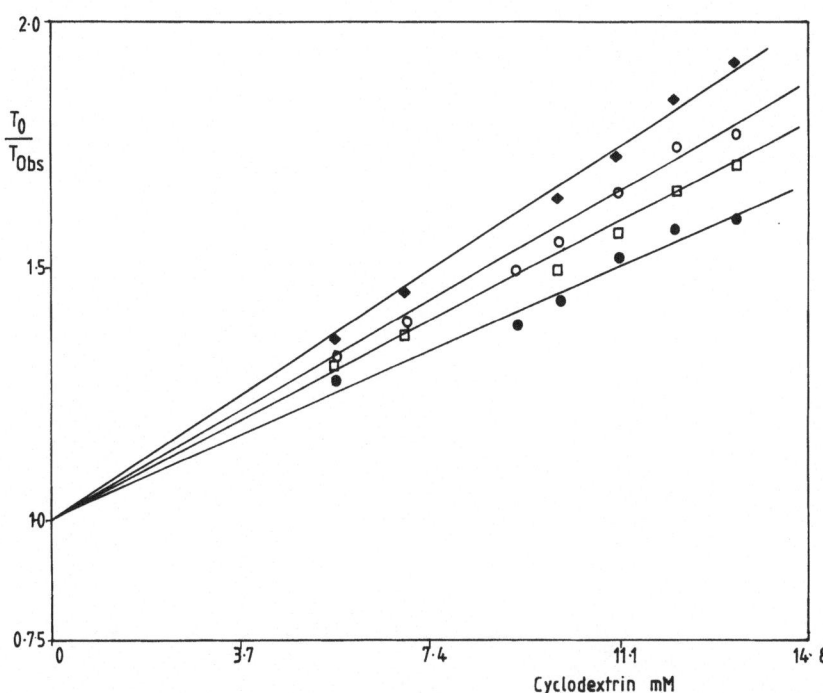

Fig. 2.   Plot of equation[2] for the determination of association constants, conditions as
Fig. 1
◆ DNS-D-Val,   ○ DNS-D-Thr,   □ DNS-L-Val,   ● DNS-L-Thr.

Table

Compound	Chromatographic K/M^{-1}	$\Delta\Delta G/J\,mol^{-1}$ 295 K	α at 12mM β–CD	Fluorescent K/M^{-1}
DNS-Gly	58 ± 14			24 ± 9
DNS-D-Thr	60 ± 6			
		600	1.13	
DNS-L-Thr	47 ± 5			64 ± 12
DNS-D-Val	71 ± 7			
		720	1.09	
DNS-L-Val	53 ± 6			58 ± 13
DNS-D-Leu	78 ± 13			
		650	1.11	
DNS-L-Leu	60 ± 13			48 ± 12

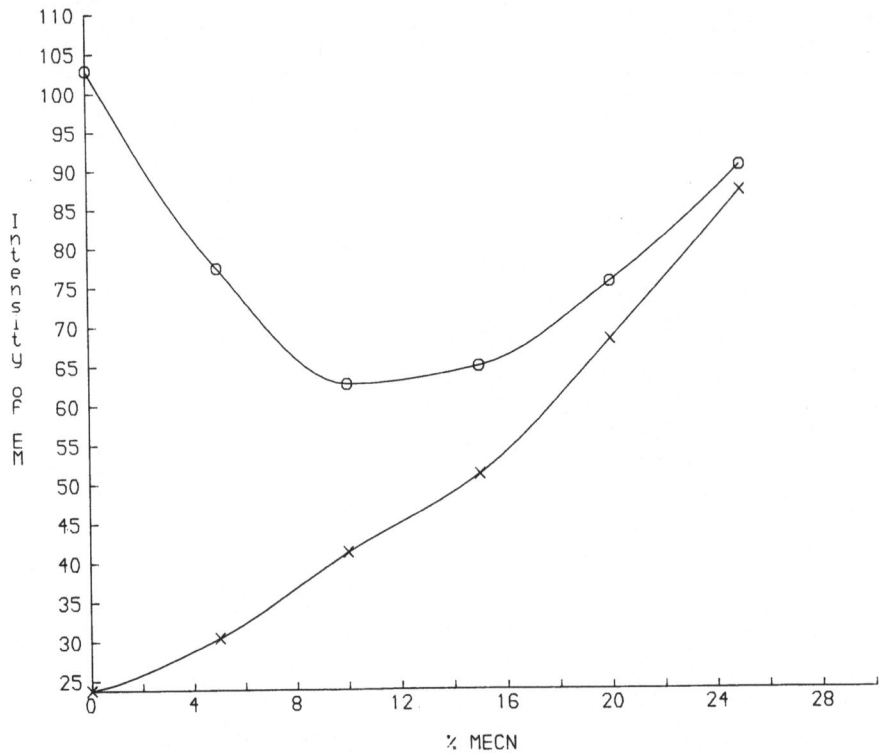

Fig. 3.   Fluorescent intensity of DNS-Gly as a function of acetonitrile concentration, in the absence of, (lower curve), and in the presence of, (upper curve), 2.5 mM β-cyclodextrin. Fluorescence excited at 338 nm and measured at 521 nm.

The chromatographic behavior of the dansyl amino acids in the presence of β-cyclodextrin has been explained in terms of a non-retained complex in which the dansyl group is included in the cyclodextrin cavity, organic solvents destabilize this complex, but D-enantiomers bind more strongly and thus have shorter retention times. The free energy differences between the two diastereomeric complexes are similar for those amino acids examined, (see table) suggesting that there is no specific interaction with functional groups on the amino acid side chains.

REFERENCES

1. T. Takeuchi, H. Asai, and D. Ishi, *J.Chromatogr.*, 357:409 (1986).
2. K. Uekama, F. Hirayama, S. Nasu, N. Matsuo and T. Irie, *Chem.Pharm.Bull.*, 26:3477 (1978).
3. R. F. Chen, *Arch.Biochem.Biophys.* 120:609 (1967).
4. T. Kinoshita, F. Iinuma and A. Tsuji, *Biochem.Biophys.Res.Comm.*, 51:666 (1973).
5. R. I. Gelb, L. M., Schwartz, M. Radeos, R. B. Edmonds and D. A. Laufer, *J. Amer. Chem. Soc.*, 104:6283 (1982).

A NOTE ON ANALYTICAL AND PREPARATIVE CHIRAL RESOLUTION OF SOME

AMINOALCOHOLS BY ION-PAIR HIGH-PERFORMANCE LIQUID CHROMATOGRAPHY

R.M. Gaskell and B. Crooks

Physical Chemistry Section
ICI Pharmaceuticals Division, Mereside
Alderley Park, Macclesfield, Cheshire SK10 4TG, UK

INTRODUCTION

Three methods of effecting chiral separations by High-Performance Liquid Chromatography (HPLC) have been used with varying degrees of success; modifications of the solute[1], the eluent[2], and the column support[3]. In our view eluent modification is currently the most flexible technique for analytical and preparative separations, because solute solubility is enhanced using chiral (ion-pair) additives, and conventional column supports may be used. Our prime objective has been the isolation of enantiomers of sufficient purity and quality for preliminary biological testing, using chiral ion-pair HPLC.

Several workers have extensively investigated chiral ion-pair chromatography. Petterson and Schill[4-6] have used a DIOL adsorbent to demonstrate selective analytical separations of several racemic aminoalcohols in a low polarity eluent containing (+) - 10-camphorsulphonate as the pairing ion.

We have achieved excellent selectivities on silica supports using eluents with increased eluting power, comprising dry isopropanol/hexane mixtures with (+) - tartaric acid as the chiral ion and pyridine as a competing base. These high polarity mobile phases were specifically developed to solvate aminoalcohol racemates in concentrations (~10 mg/ml) appropriate for preparative work.

EXPERIMENTAL

Enantiomers and mixtures of propranolol HCL (Inderal®) atenolol (Tenormin®), practolol (Eraldin®) and dimethyl-1,4-dihydro-4-(2-[4-(2-hydroxy-3-phenoxy-propy-lamino)]-5-nitrophenyl) -2,6-dimethylpyridine -3,5-dicarboxylate (M192221) were supplied 'in house'. (±)epinephrine, (±)2-benzylamino-1-propanol and (±)-α-methylbenzylamine were obtained from Aldrich (U.K.) (+) - tartaric acid and triethylamine were Analar® Grade (BDH Poole) and pyridine was Pierce silylation grade. All the mobile phase solvents were dry (HPLC Grade Fisons, Loughborough), with a controlled moisture content of <0.05%.

The analytical experiments were performed with a Varian 5560 system equipped with a variable wavelength UV detector (UV200 Varian) operated at 254 or 280 nm. Injections were made via a Varian 8085 autosampler and a 402 Data System (Varian) was used to measure the raw data. A Perkin Elmer Series 3B pump fitted with an LC 75 UV detector was used for the preparative separations. The columns used in the study were a 25 cm x 5 mm Lichrosorb DIOL (10μm), a 25 cm x 5 mm Spherisorb S5W silica, and a 25 cm x 1" Dupont Zorbax 7 μm silica preparative column all of which were available from Hichrom (U.K.). Optical rotation measurements were carried out on a Perkin Elmer 241 polarimeter at 589 nm equipped with a 1 dm cell. NMR spectra were run on a Bruker WM 400 MHz spectrometer.

®'Inderal' 'Tenormin' and 'Eraldin' are trademarks, the property of Imperial Chemical Industries PLC.

RESULTS

Analytical Scale Separations

From extensive method development, selective separations of several racemic aminoalcohols were obtained using mobile phase mixtures of isopropanol and hexane containing 0.01M(+)-tartaric acid and 0.015% triethylamine on DIOL and silica adsorbents (Fig. 1). Preliminary work was undertaken to compare the effects of different competing bases using triethylamine (pK_a ~10.7) and pyridine (pK_a ~5.1). (Table 1). The effects of varying the concentration of the competing base and the overall concentration of the (+) - tartrate - base ion-pair were also studied (Table 2 and Fig. 2).

All solutes were dissolved as hydrochlorides or free bases in isopropanol containing 0.01M (+) - tartaric acid.

Preparative Scale Separations

A specific objective of this work was to isolate the enantiomers of M192221 in sufficient quantity for biological testing. The efficiency and loading capacity of a 25 cm x 5 mm Spherisorb S5W column was compared with a similar Lichrosorb DIOL column for the separation of this racemate (Table 3). Subsequently the separation of M192221 was scaled up directly to a loading of 17 mg per pass through a DuPont Zorbax 1" x 25 cm preparative silica column (Fig. 3).

The separated isomers were recovered as free bases by base extraction and checked by NMR for levels of residual (+) - tartaric acid. Optical rotation measurements were taken in methanol after conversion to hydrochlorides (Table 4).

DISCUSSION

Ion-pair equilibria move towards dissociation as solvent polarity and the difference in pK_a between the analyte and the pairing-ion (ΔpK_a) increase [7]

$$\text{Base} + \text{H Acid} \rightleftharpoons \underset{\text{H-bonded complex}}{\text{B...HA}} \rightleftharpoons \underset{\text{H-bonded ion-pair}}{\text{BH}^+...\text{A}^-} \rightleftharpoons \text{BH}^+ + \text{A}^-$$

It was reasoned that other non-electrostatic interactions necessary for chiral recognition (e.g. H bonding) would be maximised in these associated species. To achieve this the mobile phase polarity was offset by a modest ΔpK_a ((+)-tartaric acid pK_a ~3.0, 4.5, propranolol ~8.5).

Isopropanol-hexane eluents were selected to combine solvating power and (reduced) solute- pairing ion H-bond interactions. (+) - Tartaric acid was selected as the chirally directing counter ion because of its moderate pK_a, high solubility in isopropanol, UV transparency, and its strong H-bond donor profile.

These preliminary studies have shown that the separations achieved with pyridine as a competing base are more selective than with triethylamine. High concentrations of base (>0.05%) and counter ion (>0.025M) gave inconveniently long retention times and low selectivities. Similarly low concentrations of base (<0.005%) and counter ion (<0.005M) resulted in high selectivities, poor peak efficiencies and ultimately no elution. The optimum ratio was determined as [pairing-ion]: [Base] 5:1, specifically 0.01M (+) - tartaric acid, 0.015% base.

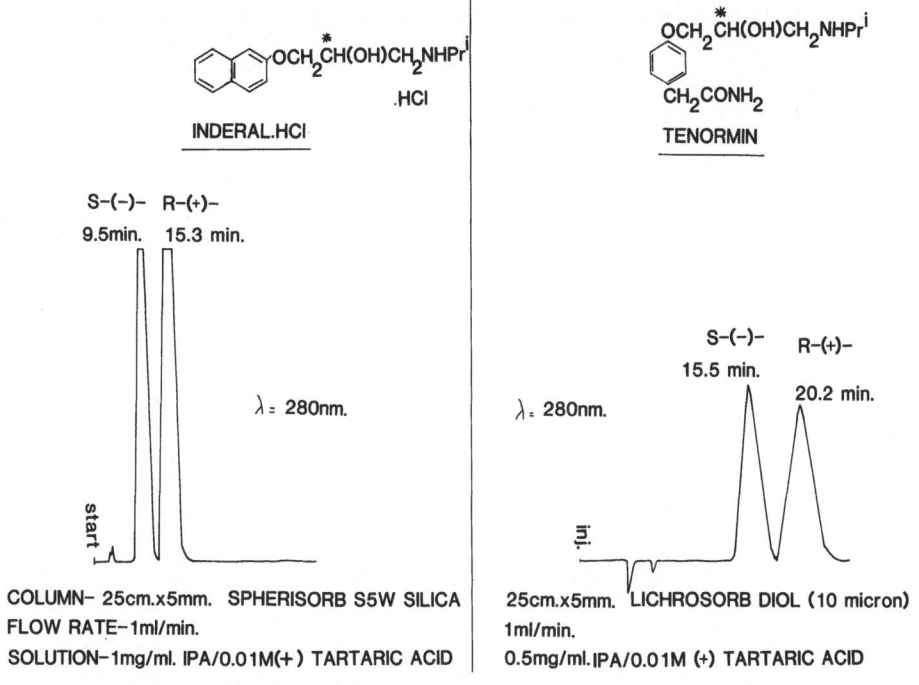

Fig. 1　Specimen chromatograms showing separation of enantiomers of Inderal and Tenormin.

Table 1.　Retention and Separation Parameters
　　　　Eluent A = 70 Isopropanol 30 hexane 0.01M(+) - tartaric acid 0.015% base.
　　　　Eluent B = 60 Isopropanol 40 hexane 0.01M(+) - tartaric acid 0.015% base.
　　　　BASE E = triethylamine pK_a ~10.7.
　　　　BASE P = pyridine pK_a ~5.1.

Phase Compound	Base	DIOL k'_1	k'_2	α	SILICA k'_1	k'_2	α	Eluent
propranolol hydrochloride	E	4.12	6.26	1.5	1.36	2.32	1.7	B
	P	-	-	-	2.24	4.19	1.9	B
atenolol	E	4.17	5.75	1.4	3.89	5.96	1.5	A
	P	-	-	-	5.83	10.27	1.8	A
practolol	E	-	-	-	1.81	2.88	1.6	A
	P	-	-	-	3.85	7.04	1.8	A
M192221	E	6.67	8.17	1.2	2.61	3.91	1.5	A
	P	6.19	8.43	1.4	6.67	11.00	1.7	B
(±) epinephrine	E	-	-	-	2.14	3.76	1.8	B
	P	-	-	-	3.37	6.97	2.1	B
(±)-2-benzylaminopropanol	E	-	-	-	2.17	2.61	1.2	A
	P	-	-	-	-	-	-	
(±) -α methylbenzylamine	E	-	-	-	1.27	1.27	1.0	B
	P	-	-	-	1.42	1.42	1.0	B

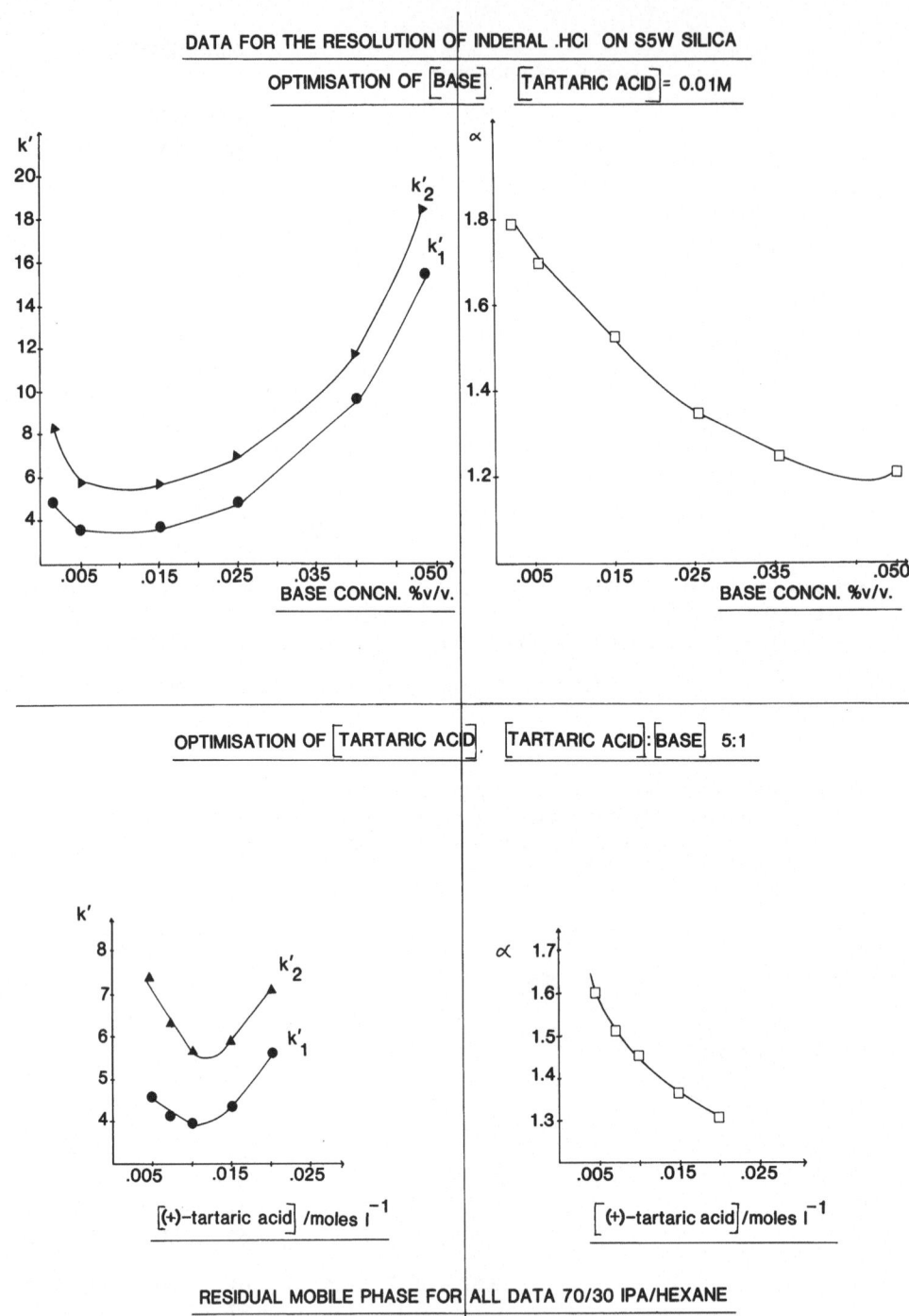

Fig. 2. Optimisation of the (a) base and (b) tartaric acid concentration for the resolution of propranolol.

It has been reported[5] that separations of some enantiomeric aminoalcohols can be achieved without a competing base. This factor coupled with our observed effects of competing base concentration on solute retention suggests a mixed separation mechanism, involving more than the adsorption of solute - tartrate ion-pairs on to silica. It is reasonable to postulate the contribution of an ion exchange mechanism in which protonated solutes exchange with the protonated competing base which has been adsorbed as an ion-pair with tartrate counter ion e.g.

$$[(+)\text{-Tartrate}^{-+}\,\text{PYR}]\text{-}[(\pm)\text{-Propranolol}^{+}] \rightleftharpoons [(+)(+)\text{-Tartrate}^{-}\,\text{Propranolol}^{+}]+\text{PYR}^{+}$$
adsorbed. $[(+)(-)\text{-Tartrate}^{-}\,\text{Propranolol}^{+}]$

Table 2.   Variation of [Base]. Data for propranolol HCl on DIOL using triethylamine

[BASE] % v/v	[(+)-Tartaric acid]	k'_1	k'_2	α
0.001	0.01M	4.63	8.25	1.8
0.015	"	3.83	5.78	1.5
0.025	"	5.03	6.74	1.3
0.035	"	7.52	9.25	1.2
0.050	"	15.38	18.25	1.2
Variation of [(+)-tartrate/base]				
0.003	0.005M	4.62	7.38	1.6
0.011	0.008M	4.07	6.15	1.5
0.015	0.01M	3.95	5.76	1.5
0.023	0.015M	4.24	5.82	1.4
0.030	0.020M	5.34	6.96	1.3

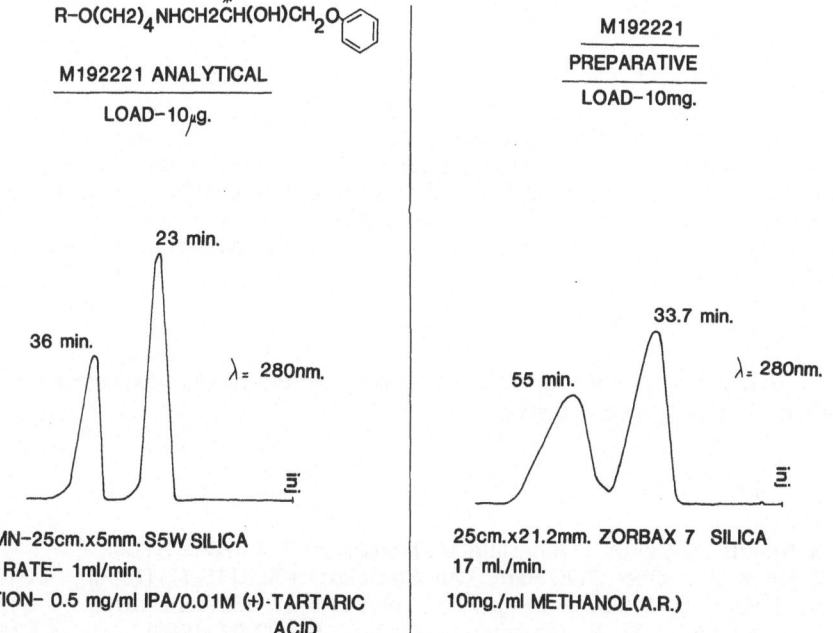

Fig. 3.   Specimen chromatograms showing analytical and preparative separation of M192221.

Table 3. Loading Capacity for M 192221

Column	Efficiency(N)	k'_1	k'_2	α	Max. load mg	Eluent
25cm x 5mm Lichrosorb Diol (10μ)	10,000	6.19	8.43	1.4	0.2	B
25cm x 5mm Spherisorb S5W	20,000	6.67	11.00	1.7	1.0	B
Zorbax 1" x 25cm Silica	16,000	5.50	9.58	1.7	17.0	B

Table 4. Analysis of Isolated Enantiomers from M192221

Racemate	Enantiomer	Chiral Purity (HPLC)	(+)-Tartaric -Acid (NMR)	$[\alpha]_D^{RT}$
M192221	(-)-	> 95%	< 2%	-11.07
	(+)-	> 95%	< 2%	+11.58

Underivatised silica has more adsorption sites than comparable bonded phase materials and promotes a faster more efficient mass transfer process. Consequently it was anticipated that silica supports would enhance the loading capacity and column selectivity compared with DIOL. This was conclusively demonstrated for all the separations studied.

It was observed that the (-) isomers eluted first in these systems indicating a preferential solvation of this enantiomer in the chiral mobile phase.

CONCLUSIONS

The systems described have demonstrated excellent selectivities for a range of aminoalcohol racemates. The preparative potential of the method has been exploited for the isolation of 10-20 mg of racemic material of sufficient quality for biological testing. The method has also been used to measure the optical purity of chemically resolved material down to levels of 0.5% of one enantiomer in the 'pure' enantiomer (unpublished observations).

Acknowledgement

We thank Mr. J. Coope for technical assistance and helpful discussions and Mrs. D. Rawsthorn for typing the manuscript.

REFERENCES

1. N. Nimura, H. Ogura, T. Kinoshita, *J.Chromatogr.*, 202:375-9, (1980).
2. E. Gil-av, A. Tishbee, P. E. Hare, *J.Am.Chem.Soc.*, 102:5115-17, (1980).
3. W. H. Pirkle, D. W. House, *J.Org.Chem.*, Vol. 44 No. 12, 1957-60, (1979).
4. C. Pettersson, G. Schill, *Chromatographia* Vol. 16, 192-97 (1982).
5. C. Pettersson, G. Schill, *J.Chromatogr.*, 204, 179-83, (1981).
6. C. Pettersson, G. Schill, *J.Chromatogr.*, 316, 553-67 (1984).
7. H. Ratajczak, W. J. Orville-Thomas "Molecular Interactions" Vol. 2, 179-229 (1981).

SIMULTANEOUS ENANTIOSELECTIVE DETERMINATION OF UNDERIVATIZED

β-BLOCKING AGENTS AND THEIR METABOLITES IN BIOLOGICAL SAMPLES

BY CHIRAL ION-PAIRING HIGH-PERFORMANCE LIQUID CHROMATOGRAPHY

T. Leemann and P. Dayer

Clinical Pharmacology Unit, University Hospital
24, Rue Micheli-du-Crest, 1211 Geneva 4, Switzerland

SUMMARY

Simultaneous enantioselective determination of various β-blocking agents and their metabolites without prior derivatization is made possible by the technique presented. Resolution of enantiomers is achieved by use of a chiral pairing-ion in a non-aqueous mobile phase and elution on a standard non-chiral column. Applicability and performance are illustrated by results obtained for metoprolol and its two major oxidised metabolites, both *in vivo* (blood and urine) and *in vitro* (human liver microsomes incubations). Baseline separation of all compounds is obtained in less than 20 minutes. The technique is reliable and shows broad applicability to studies involving β-blocking agents and possibly other structurally related compounds.

INTRODUCTION

Stereochemical aspects of drug-biological systems interactions show both clinical and investigational relevance in many areas of pharmacology and toxicology and are attracting rapidly growing interest[1]. Most β-adrenoceptor blocking agents are available for therapy as racemic mixtures, although use of pure enantiomers may significantly improve their desired/unwanted effects ratio[2-6].

To investigate stereoselectivity in the pharmacokinetics of β-blockers some authors have used an indirect approach[7-9]: Here the racemate, then one[8] or both enantiomers[7,9] are administered on separate occasions to the same subjects, and classical non-stereoselective assays are used for quantitation. Such an approach suffers major drawbacks, both ethical and methodological: the same subjects have to be exposed several times to prototype compounds, day to day intra-individual variability may lead to erroneous or insufficient results, pharmacokinetic or pharmacodynamic interactions between enantiomers may further complicate interpretation, and racemization or stereoselective inversion[10-11] cannot be detected.

In the past decade several methods have also been developed for the analytical resolution of optical isomers of propranolol, metoprolol or other β-blocking agents[12-33], thus allowing a more direct approach that potentially obviates the above-mentioned problems. Published techniques include:

(a)   stereoselective immunoassay,

(b)   gas chromatographic separations on achiral stationary phases of either "pseudoracemates" with mass-spectrometric detection or diastereomeric derivatives formed with optically pure chiral reagents

(c)   liquid chromatography on achiral stationary phases, either after derivatization to diastereoisomers, or by ion-pairing with an optically pure chiral pairing-ion dissolved in the mobile phase

(d)   liquid chromatography on chiral stationary phases.

As our research interest is in the elucidation of stereochemical factors involved in the genetically determined metabolic oxidation of various β-blockers, both *in vivo* and *in vitro*, and their incidence on pharmacokinetic-pharmacodynamic relationships, then simultaneous quantification of the isomers of both drug and metabolite isomers and sensitivity of a few nanograms (5-50 pmoles) or lower are prerequisites. Processing ease and speed, use of standard equipment, preferably HPLC, as well as versatility and overall cost are also important considerations. The ion-pairing approach proposed by Pettersson et al,[19] seemed most promising with respect to these constraints.

The interest and performance of this approach in metabolic studies both *in vivo* and *in vitro* is illustrated here by results from the development and application of the technique for metoprolol and its two major oxidised metabolites using a new chiral pairing-ion[33].

EXPERIMENTAL

Materials

The eluent was delivered by a Spectroflow 400 pump (Kratos, Westwood, NJ, USA) and a ISS-100 autosampler (Perkin Elmer, Norwalk, CT, USA) with a 150 μl loop used for injection. Separations were carried out on a Lichrosorb Diol CGC glass cartridge column (150 x 3.2 mm I.D., 5 μm irregular particles) from E. Merck (Darmstadt, FRG). A Schoeffel FS-970 L.C. Spectrofluorometer (Kratos, Westwood, NJ, USA) with a 5 ml cell and a SP 4290 integrator (Spectra-Physics, San Jose, CA, USA) were used for detection and integration of eluting peaks.

Chemicals

Racemic metoprolol tartrate, R-(+)-metoprolol hydrochloride (H 150/64), S-(-)-metoprolol tartrate (H 150/65) and racemates of the oxidised metabolites 4'-hydroxy-metoprolol p-hydroxy-benzoate (H 119/66) and O-desmethyl-metoprolol p-hydroxybenzoate (H 105/22) were donated by Hässle (Dr. C.-G. Regardh, Mölndal, Sweden). Racemic 4'-hydroxy-betaxolol fumarate, donated by L.E.R.S-Synthelabo (Paris, France), was used as internal standard. Triethylamine (TEA), analytical grade dichloromethane and methanol were purchased from E. Merck. N-benzoxycarbonyl-glycyl-L-proline (ZGP) and Dehydrat molecular sieves were obtained from Fluka (Buchs, Switzerland). Extrelut® 3 (E. Merck) and 100 mg 1 ml BondElut® C18 (Analytichem International, Harbor City, CA, USA) columns were used for extraction.

Standard Solutions

Methanolic solutions of each compound were prepared and further diluted in 10^{-5} M HCl or dichloromethane as needed. These solutions were kept in the refrigerator at <4°C and appeared to be stable for several months under these conditions.

Chromatographic Conditions

The mobile phase, consisting of methanol 1.1%, ZGP 6.2 mM and triethylamine 300 μM in dichloromethane, was dried on molecular sieves. A flow rate of 0.6 ml/min was used.

The excitation wavelength was set at 230 nm and a CS-7-54 band filter (maximum ~320 nm) used for emitted light.

Enantiomer Elution Order

Enantiomeric identity of eluting peaks of metoprolol was assigned by direct injection of pure isomers. For the metabolites, isomeric identity was determined by incubation of pure metoprolol enantiomers with human liver microsomal preparations[34].

Extraction

Blood: 1 to 2 ml samples, aqueous internal standard and pH 10, 1M carbonate buffer were successively added (up to a 3.7 ml total volume) to Extrelut® columns; columns were eluted with 2 x 5 µl 5% methanol in dichloromethane; the eluate was then evaporated to dryness at 40°C under a stream of nitrogen; dried extracts were redissolved in dichloromethane (300 µl) and left overnight to evaporate in open 300 µl autosampler vials; 170 µl of mobile phase was then added and 150 µl aliquots injected into the chromatographic system.

Incubations: 400 ml incubations were extracted in exactly the same way as blood samples.

Urine: 400 µl pH 10 1 M carbonate buffer, 200-500 µl samples and aqueous internal standard were added to BondElut® columns; after vacuum suction, columns were successively rinsed with distilled water (400 µl) and 40% methanol in water (400 µl); compounds of interest were then eluted by 0.1% ammonia in 1:1 methanol/acetonitrile (2 x 2 ml); evaporation and injection were performed as described for blood.

Standard Plot and Variability

Blood (2ml) and urine (200 µl) samples were spiked with 15, 30, 45, 75, 105, 150, 225, 450, 750, 1500, 2250 pmoles (~5 to 750 ng) of each metoprolol isomer and 25, 50, 75, 125, 175, 250, 375, 750, 1250, 2500, 3750 pmoles (~10 to 1500 ng) of each metabolite isomer.

Intra-day variability for blood samples was estimated from 6 replicate samples at 15, 52.5, 375 pmol/ml (~5, 17.5, 125 ng/ml) and 10, 105, 750 pmol/ml (~10,35,250 ng/ml) of metoprolol and the metabolite isomers respectively. In urine variability was estimated using 10 fold higher concentrations.

RESULTS AND DISCUSSION

Chromatographic Conditions

Methanol, instead of 1-pentanol[33], was used as a mobile phase modifier, since we observed that aliphatic alcohols of decreasing alkyl chain length could be used in lower concentrations and gave shorter retention times while maintaining separation efficiency, thus increasing sensitivity.

In the range of conditions tested the R-(+)-enantiomer eluted first in all cases (Fig. 1). The elution order of the internal standard enantiomers was not checked directly, but is assumed to be the same as for metoprolol and its metabolites: its chemical structure is very similar to that of 4'-hydroxy-metoprolol, and R-(+)-betaxolol (tested in our laboratory) also elutes first. The same elution order has been reported for alprenolol and propranolol enantiomers as well[33], and it is consistant with results previously obtained with D-(+)-10-camphorsulfonic acid as the pairing-ion[19,26].

The effects of varying pairing-ion concentrations (2.5 to 7.2 mM), methanol (0.9 to 1.2%) and triethylamine (1 to 900 mM) on retention times and selectivity can be summarized as follows:

- increasing methanol concentrations shortens retention times, increases sensitivity, but reduces stereoselectivity while improving separation of enantiomeric pairs from one another (Fig. 2A-C);

- in the range tested, increasing triethylamine concentrations had hardly any effect on overall retention times and sensitivity, but reduced stereoselectivity while improving separation of enantiomeric pairs from one another (Fig. 2D-E);

- increasing pairing-ion concentration shortens retention times, tends to reduce sensitivity by increasing background fluorescence, and increases stereoselectivity while favoring overlaps of S-(-)-hydroxy-betaxolol with R-(+)-hydroxy-metoprolol and S-(-)-hydroxy-metoprolol with R-(+)-O-desmethyl-metoprolol (Fig. 2, E-F);

The final working eluent composition (see Experimental) offered a compromise between these counteracting effects and was arrived at by optimization using the basic "one-factor-at-a-time" method. Baseline separation was obtained in less than 20 minutes (Fig. 3), with separation factors of 1.16 for each enantiomeric pair.

Extraction and Quantification

Extraction coefficients from blood, urine or incubations were above 90% for all compounds. The presence of methanol in the dichloromethane was required for high extraction yields from Extrelut® columns whilst recovery from BondElut® columns was much improved by the presence of ammonia in the eluting solvent.

Measurement limits were around 10 and 25 pmoles (~2.5 and 7 ng) injected metoprolol and metabolite enantiomers, respectively. Linearity was observed over a very wide range for directly injected material. In addition linearity and good correlations (Table 1) were also observed after extraction on both Extrelut® (blood standard plots) and Bondelut® (urine standard areas) columns, with intra-day coefficients of variation below 10% for the concentrations tested.

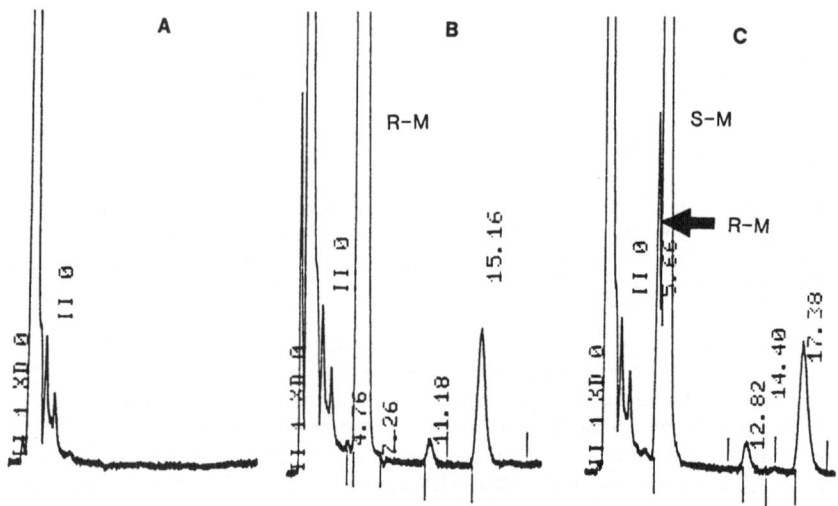

Fig. 1. *In vitro* production of metabolites from metoprolol isomer incubation (human liver microsomes). A) Blank sample. B) R-metabolites (11.18 and 15.16 min) from R-metroprolol (R-M). C) S-metabolites (12.82 and 17.38 min) from S-metoprolol (SM-M). Chromatographic conditions: see Experimental.

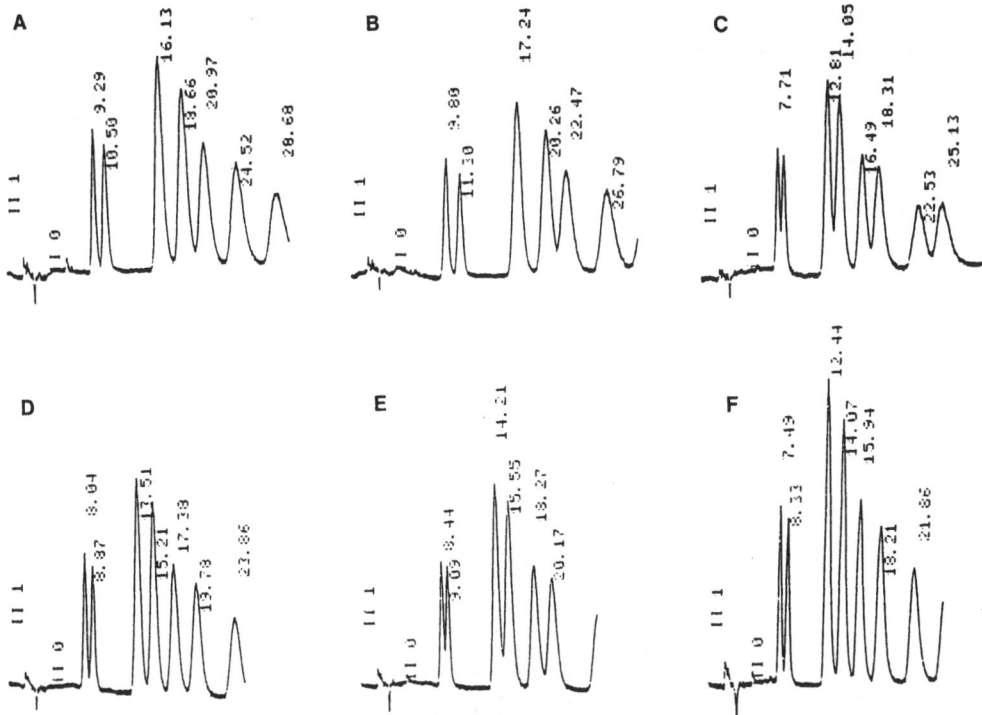

Fig. 2. Influence of eluent composition. A) ZGP 2.6 mM, methanol 1.0% B) ZGP 2.6 mM, methanol 0.9% C) ZGP 2.6 mM, methanol 1.2% D) Methanol 1%, ZGP 3.2 mM, TEA 300 μM E) Methanol 1%, idem, TEA 400 mM F) Methanol 1% ZGP 3.9 mM, idem. Direct injection of metoprolol, internal standard and metabolites racemates (0.3, 0.5, 0.5, 0.5 nmoles, resp.) in dichloromethane/methanol (99.5:0.5).

Applications

Previous application of the ion-pairing technique with D-(+)-10-camphorsulfonic acid as pairing-ion had already proven useful in metabolic studies involving β-blocking agents, both *in vivo* and *in vitro*[35-38]. Genetically determined mono-oxygenase activity of the debrisoquine type, for example, is highly stereoselective in humans and is lacking in about 10% of so-called "poor metabolizers" in Western White populations, as illustrated in Fig. 4. Selective inhibitors with very high affinity for the genetically controlled isozyme are known (for example, quinidine,) and efficient stereoselective quantitative techniques such as described here are key tools in metabolic interaction studies[37].

In vivo, both in blood and urine, only 4'-hydroxy-metoprolol can be detected, as O-desmethyl-metoprolol is rapidly further oxidised at its terminal hydroxyl group to a carboxylic acid. This latter metabolite can also be chirally resolved on the same chromatogram, but cannot be extracted by the procedures described. *In vitro* both 4'-hydroxy- and O-desmethylmetoprolol can be quantified.

Interestingly, oxidations to both 4'-hydroxy- and O-desmethyl-metoprolol are under the same genetic control, but with opposite stereoselectivity[38], a pehnomenon also observed with bufuralol 4- and 1'-hydroxylations[36]. This observation further points to the need for efficient methodology to investigate chiral factors involved in drug metabolism.

The use of Z-glycyl-L-proline as pairing-ion brings a clear improvement in selectivity, retention times and sensitivity over D-(+)-10-camphorsulfonic acid. The technique may

Table 1. Correlation Coefficients for Blood and Urine Standard Plots

| | Blood (2ml) | | Urine (200 μl) | |
	R-(+)	S-(-)	R-(+)	S-(-)
Metroprolol	1.000*	1.000*	0.997	0.997
4'-Hydroxy-metoprolol	1.000*	1.000*	0.999	0.998
O-Desmethyl-metoprolol	1.000*	1.000*	0.998	0.998

* >0.9995

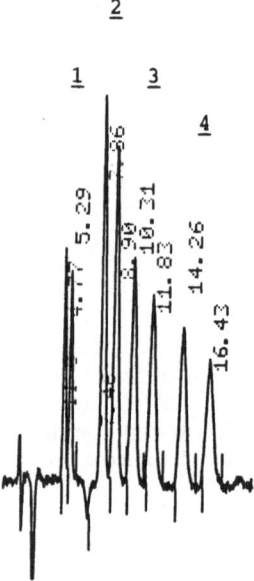

Fig. 3.   Simultaneous chiral resolution of metoprolol and metabolites (direct injection as in Fig. 1). *1*. R- and S-metoprolol *2*. R- and S-4'-hydroxy-betaxolol *3*. R- and S-4'-hydroxy-metoprolol *4*. R- and S-O-desmethyl-metoprolol.

also be suitable for a wider range of substances: oxprenolol isomers, for example, can be resolved using Z-glycyl-L-proline, whereas no separation is obtained with D-(+)-10-camphorsulfonic acid[19].   Table 2 lists some β-blocking agents and their oxidised metabolites to which ion-pairing with Z-glycyl-L-proline can be applied, sometimes after slight modifications in detection, chromatographic or extraction conditions, for qualitative or quantitative studies both *in vivo* and *in vitro*. This list is only indicative and other structurally related drugs or metabolites may and will probably be added with time.

Chiral ion-pairing HPLC is indeed a versatile approach and can certainly be just as easily and efficiently implemented for studies involving analytical resolution of acidic chiral drugs[39-41].

CONCLUSIONS

Ion-pairing HPLC using Z-glycyl-L-proline as a chiral counter-ion is a straightforward, efficient and reliable stereoselective technique that shows broad applicability to studies

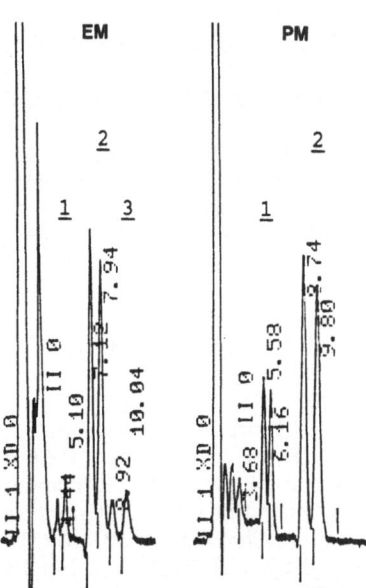

Fig. 4. Six hour blood profiles in an extensive (EM) and a poor (PM) metabolizer human volunteers (debrisoquine-type of oxidation polymorphism) after metoprolol 50 mg single oral dose.
1) Metoprolol - 2) 4'-hydroxy-betaxolol (I.S.) - 3) 4'-hydroxy-metoprolol.

Table 2. Some β-Blockers and Their Metabolites that can be Resolved and Quantified by Ion-Pairing HPLC with ZGP

Alprenolol	Oxprenolol
4-Hydroxy-alprenolol	Propafenone
Betaxolol	5-Hydroxy-propafenone
4'-Hydroxy-betaxolol	Propranolol
Burfuralol	4-Hydroxy-propranolol
1'-Hydroxy-bufuralol	
Metoprolol	
4'-Hydroxy-metoprolol	
O-Desmethyl-metoprolol	

involving β-blocking agents (see also Gaskell and Crooks this vol.) and possibly other structurally related compounds. Moreover simultaneous quantification of drugs and metabolites is possible, both *in vivo* and *in vitro*.

Acknowledgement

Part of this work was supported by the Fondation de Recherches Medicales C. & E. de Reuter - Geneva.

REFERENCES

1. E. J. Ariens, W. Soudijn, P. B. M. W. M. Timmermans eds., "Stereochemistry and Biological Activity of Drugs," Blackwell Scientific Publications, Oxford, (1983).
2. M. G. Myers and H. F. Hope-Gill, *Clin.Pharmacol.Ther.*, 25:303 (1979).
3. P. Heyma, R. G. Larkins, L. Higginbotham and K. Wah Ng, *Br.Med.J.*, 2:24 (1980).

4. G. D. Johnston, M. B. Finch, J. A. McNeill and R. G. Shanks, *Br.J.Clin.Pharmacol.*, 20:507 (1985).

5. R. A. Reeves, G. L. A., From, W. Paul and F. H. H. Leenen, *Clin.Pharmacol.Ther.*, 37:157 (1985).

6. R. Richards and A. E. Tattersfield, *Br.J.Clin.Pharmacol.*, 20:459 (1985).

7. R. J. Francis, P. B. East and J. Larman, *Eur.J.Clin.Pharmacol.*, 23:529 (1982).

8. C. Harvengt and J. P. Desager, *Int.J.Clin.Pharmacol.Ther.Toxicol.*, 20:57 (1982).

9. C. F. George, T. Fenyvesi, M. E. Conolly and C. T. Dollery, *Eur.J.Clin.Pharmacol.*, 4:74 (1976).

10. W. J. Wechter, D. G. Loughead, R. J. Reischer, G. J. Van Giessen and D. G. Kaiser, *Biochem.Biophys.Res.Comm.*, 61:833 (1974).

11. A. Rubin, M. P. Knadler, P. P. K. Ho, L. D. Bechtol and R. L. Wolen, *J.Pharm.Sci.*, 74:82 (1985).

12. H. Ehrsson, *J.Pharm.Pharmacol.*, 28:662 (1976).

13. K. Kawashima, A. Levy and S. Spector, *J.Pharmacol.Exp.Ther.*, 196:517 (1976).

14. S. Caccia, C. Chiabrando, P. De Ponte and R. Fanelli, *J.Chromatogr.Sci.*, 16:543 (1978).

15. S. Caccia, G. Guiso, M. Ballabio and P. De Ponte, *J.Chromatogr.*, 172:457 (1979).

16. T. Walle and U. K. Walle, *Res.Comm.Chem.Pathol.Pharmacol.*, 23:453 (1979).

17. J. Hermansson and C. Von Bahr, *J.Chromatogr.*, 221:109 (1980).

18. B. Silber and S. Riegelman, *J.Pharmacol.Exp.Ther.*, 215:643 (1980).

19. C. Pettersson and G. Schill, *J.Chromatogr.*, 204:179 (1981).

20. W. H. Pirkle, J. M. Finn, J. L. Schreiner and B. C. Hamper, *J.Am.Chem.Soc.*, 103:3964 (1981).

21. J. Hermansson, *Acta.Pharm.Suec.*, 19:11 (1982).

22. J. Hermansson and C. Von Bahr, *J.Chromatogr.*, 227:113 (1982).

23. J. A. Thompson, J. L. Holtzman, M. Tsuru, C. L. Lerman and J. L. Holtzman, *J.Chromatogr.*, 238:470 (1982).

24. T. Walle, M. J. Wilson, U. K. Walle and S. A. Bai, *Drug Metab.Disp.*, 11:544 (1983).

25. A. J. Sedman and J. Gal, *J.Chromatogr.*, 278:199 (1983).

26. T. Leemann, P. Dayer, L. Balant and P. Buri, *Experientia.*, 40:647 (1984).

27. M. G. Sankey, A. Gulaid and C. M. Kaye, *J.Pharm.Pharmacol.*, 36:276 (1984).

28. I. W. Wainer, T. D. Doyle, K. H. Donn and J. R. Powell, *J.Chromatogr.*, 306:405 (1984).

29. J. W. Wilson and T. Walle, *J.Chromatogr.*, 310:424 (1984).

30. D. W. Armstrong, T. J. Ward, R. D., Armstrong and T. E. Beesley, *Science*, 232:1132 (1986).

31. A. Darmon and J. P. Thenot, *J.Chromatogr.*, 374:321 (1986).

32. P. H. Hsyu and K. M. Giacomini, *J.Pharm.Sci.*, 75:601 (1986).

33. C. Pettersson and M. Josefsson, *Chromatographia*, 21:321 (1986).

34. P. Dayer, T. Kronbach, M. Eichelbaum and U. A. Meyer, *Biochem.Pharmacol.*, 36:4145 (1987).

35. P. Dayer, T. Leemann, J. Gut, T. Kronbach, A. Kuepfer, R. Francis and U. A. Meyer, *Biochem.Pharmacol.*, 34:399 (1985).

36. P. Dayer, T. Leemann, A. Kuepfer, T. Kronbach and U. A. Meyer, *Eur.J.Clin.Pharmacol.*, 31:313 (1986).

37. T. Leemann, P. Dayer and U. A. Meyer, *Eur.J.Clin.Pharmacol.*, 29:739 (1986).

38. J. Gut, R. Gasser, P. Dayer, T. Kronbach, T. Catin and U. A. Meyer, *FEBS Lett.*, 173:287 (1984).

39. C. Pettersson and G. Schill, *Chromatographia.*, 16:192 (1982).

40. C. Pettersson and K. No, *J.Chromatogr.*, 282:671 (1983).

41. C. Pettersson, *J.Chromatogr.*, 316:553 (1984).

CHIRAL GAS AND HIGH-PERFORMANCE LIQUID CHROMATOGRAPHIC ANALYSIS

OF ENANTIOMERS OF FUNGICIDES AND PLANT GROWTH REGULATORS:

APPLICATION IN FUNGAL, PLANT AND SOIL METABOLISM STUDIES

T. Clark, A.H.B. Deas and K. Vogeler*

Department of Agricultural Sciences, University of Bristol
Institute of Arable Crops Research, Long Ashton Research
Station, Long Ashton, Bristol BS18 9AF, UK and
*Bayer AG, Metabolism Institute, Pflanzenschutz Zentrum
Monheim, D-5090 Leverkusen, FRG

ABSTRACT

Chiral gas and high performance liquid chromatography columns were evaluated for the analytical (GC and HPLC) separation of enantiomers of a range of azole, pyrimidine and morpholine fungicides and plant growth regulators. Only hydroxy 1,2,4-triazoles were separated sufficiently to be of practical use. Chromatograms of the best separations obtained are shown. Of the two GC phases evaluated only "Chirasil Val" was able to separate the enantiomers of the hydroxy triazoles. Of the three cyclodextrin HPLC phases available the α column was of no use whereas both the β and β acetylated columns gave separations good enough to be used practically. Chiral GC using "Chirasil Val" showed that when triadimefon (1R + 1S) was supplied to a range of fungi it was converted primarily to triadimenol and that each fungus produced different proportions of the four possible enantiomers. A similar conversion was observed in barley plants but the same proportions of triadimenol enantiomers were formed in each of the three cultivars examined.

INTRODUCTION

Many of the agrochemicals introduced over the last 20 years possess chiral centres but are usually sold as racemic mixtures or even as mixtures of diastereoisomers. When individual enantiomers have been isolated they often display large differences in the nature and degree of their biological activity[1]. This is well documented for the important azole, pyrimidine and morpholine fungicides which became commercially available in the 1970's. For example, the 1S, 2R enantiomer of the systemic fungicide triadimenol, [1-(4-chlorophenoxy)-3,3-dimethyl-1-(1,2,4-triazol-1-yl)butan-2-ol], has been reported[2] to be 1000 times more active against the fungus *Gaeumannomyces graminis* than the 1S,2S form. Similarly the 3R enantiomer of the fungicide dichlopentezol (S-3308) is over 100 times more fungicidally active than the 3S enantiomer, the latter possessing marked plant growth regulatory properties[3]. With the plant growth regulator (PGR) paclobutrazol [(2RS, 3RS)-1-(4-chlorophenyl)-4,4-dimethyl-2-(1,2,4-triazol-1-yl)pentan-3-ol] the situation is more complex. The plant growth regulatory activity is produced by the 2S,3S enantiomer which inhibits gibberellin biosynthesis[4,5] the 2R,3R form is fungicidal[4]. However, recently we have shown[6] that the 2RS,3SR diastereoisomer of paclobutrazol is a potent inhibitor of plant sterol biosynthesis. Therefore, the ability to separate enantiomers of fungicides and PGRs is of great importance in agricultural research, particularly in studies of their biological activity, mode of action, metabolism and persistence in the environment.

Although manufacturers are beginning to release information regarding the biological activity of individual enantiomers of their products, and occasionally the synthetic routes by which they were obtained, there is a paucity of information regarding the chromatographic methods employed to determine chiral purity and stability.

We have evaluated a number of different chiral GC and HPLC columns for the analytical separation of enantiomers of a range of azole, pyrimidine and morpholine fungicides and PGRs (structures see Fig. 1). Successful separations are reported in this paper and examples are given of their application in studies of metabolism, persistence and stability of enantiomers in aqueous buffer, fungal liquid cultures, plants and soil.

EXPERIMENTAL

Materials

All agrochemicals (see Fig. 1 for structures) were kindly donated by the manufacturers or are available commercially.

The sample of bitertanol was predominantly the 1RS, 2SR diastereoisomer. Paclobutrazol comprised ca. 95% of the 2RS,3RS diastereoisomer the remainder being the 2RS,3SR form. The individual enantiomers of triadimenol, paclobutrazol and diclobutrazol were either supplied by the manufacturers or were prepared in our laboratory. The "Chirasil Val" and RSL-007 columns were supplied by Alltech Associates. The β, β-acetylated and α-cyclodextrin columns were kindly loaned by Technicol. The "Pirkle" columns were the (S)-N-3,5-dinitrobenzoylleucine type purchased from Baker (No. 71150).

Chromatography

GC analyses (using FID or ECD) were carried out on "Chirasil Val" (25 m x 0.24 mm I.D.) or RSL-007 (25 m x 0.25 mm I.D.) columns using a split/spitless injector in the spitless mode, with hydrogen as carrier gas (10 psi). Flow rates and temperature conditions were altered to optimise resolution of the enantiomers.

In the HPLC analyses, the cyclodextrin columns (25 cm x 4.6 mm I.D.) were operated isocratically in a reversed-phase mode using methanol-water or acetonitrile-water mixtures, varying the proportions of water and the flow rates to optimise the separations. The "Pirkle" columns (6 in series at 80°C 25 cm x 4.6 mm I.D.) were eluted isocratically (1 ml/min.) with hexane-dichloromethane-2-propen-1-ol (94:5:1 by volume).

Chiral SFC separations were achieved using a β-cyclodextrin column (25 cm x 4.6 mm I.D.) with CO_2 as carrier and methanol as modifier.

Metabolism of Fungicides

(i) *Studies with individual enantiomers.* Details of the methods used for the stability studies of individual enantiomers in buffer solutions have already been reported[2,7] as have those used for the fungal liquid culture[2,7-11] and plant metabolism[12] work.

Soil samples (200g) were refluxed with methanol: 0.88 ammonia (70:30 by volume, 3 x 250 mls) for 15 hrs. The extracts were reduced in volume and the triadimenol fraction purified by column chromatography and reversed phase HPLC as described elsewhere for the plant study[12].

(ii) *Triadimefon Studies.* 14C-labelled triadimefon was applied to the last fully emerged leaf of barley plants six weeks after sowing. To each leaf 100 x 0.3 μl drops (9.1×10^{-2} μCi) of 0.1 g l^{-1} Triton X-114 aqueous solution saturated with radiolabelled triadimefon were randomly spotted using a Burkard micro-applicator. After 7 and 21 days leaves were harvested, dipped in acetone (1 min) and then macerated in methanol-water-0.88 ammonia (65:23:12 by volume). A detailed description of the work up procedures will be published elsewhere.

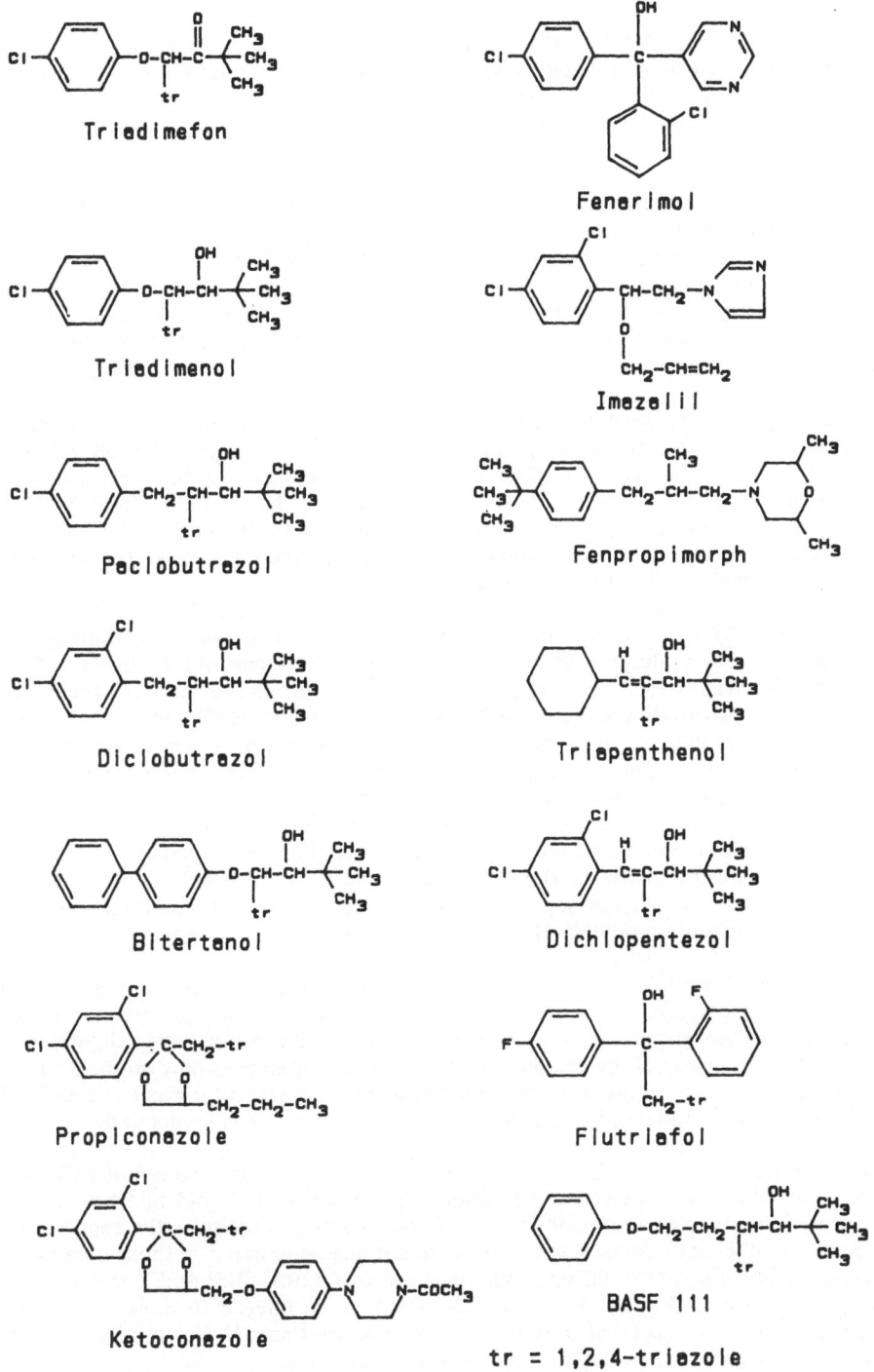

Fig. 1.   Structures of some azole, pyrimidine and morpholine fungicides and PGRs

RESULTS AND DISCUSSION

Chiral Analysis

GC on "Chirasil Val" and RSL-007. Due to the low thermal stability of "Chirasil Val" (210°C maximum operating temperature) some compounds, such as ketaconazole, were not eluted from the column even after 2 hours. Compounds which possessed a keto group rather than a hydroxyl group, such as triadimefon, could not be resolved on this phase. Chromatograms of compounds where the best enantiomer resolutions were obtained are shown in Fig. 2. Complete resolution was obtained for the enantiomers of triadimenol and of paclobutrazol. These two compounds are structurally very similar, the only difference being that the oxygen in the former is replaced by methylene in the latter. However, when a second chlorine atom is introduced into the aromatic ring of paclobutrazol forming diclobutrazol the resolution was significantly decreased. The degree of separation obtained for the enantiomers of diclobutrazol was insufficient to be of practical use. Similarly when the chlorine atom in the aromatic ring of triadimenol was replaced by a phenyl group the corresponding enantiomers were poorly resolved from one another.

Of the compounds we have analyzed only the hydroxy triazoles could be separated into their enantiomers (Fig. 3.) Those compounds successfully resolved by "Chirasil Val" were also analyzed by an RSL-007 column. None of the compounds was separated by this column even though for both phases the chiral moiety was *L*-valine-*tert*-butylamide. The two phases possess different polymeric backbones, which probably lead to different conformations of the chiral moiety, and this could well change the specific interactions between the solute and stationary phase which would in turn affect resolution.

HPLC on Cyclodextrins. Three different types of cyclodxetrin columns, namely β, β-acetylated and α, were evaluated for the separation of various pesticides. In virtually all cases the α column was inferior to the β and β-acetylated columns. As for the "Chirasil Val" GC column only compounds containing a hydroxyl group were significantly resolved into their constituent enantiomers which pointed to some involvement of hydrogen bonding in the "recognition process". For triadimenol the best resolution was obtained using the β-acetylated cyclodextrin column (Fig.4) although the enantiomers of the 1SR, 2RS diastereoisomer were only partially resolved. However, on the β-acetylated column complete enantiomer resolution was achieved for triapenthenol (Fig. 4) and this method could well be used to isolate small quantities of the individual enantiomers as well as for analytical purposes. For diclobutrazol good separations were obtained on both columns with the β column giving a clear separation (Fig. 4).

In the case of flutriafol good resolution of enantiomers was obtained on the β column whereas on the β-acetylated column the separation was inferior. Paclobutrazol was separated into its four enantiomers by both columns, the β column being slightly more efficient (Fig. 4). Resolution of the compound BASF-111 was attempted only on the β column but a good separation of enantiomers was obtained (Fig.4). In general it was not possible to predict which column would give the best separation for a given type of compound.

The β column was also evaluated using super critical fluid chromatography (SFC) for triadimenol, diclobutrazol, paclobutrazol, triapenthenol, BASF 111 and bitertanol. Flow rate (i.e. pressure), modifier and modifier content were altered to optimise the separation of enantiomers. In all cases inferior resolutions were obtained compared to those obtained by conventional reversed-phase HPLC and only paclobutrazol (2RS, 3RS) and BASF 111 were separated to any degree (Fig.5). Although separations were inferior to those obtained by conventional HPLC it is important to note that the analysis times (3-5 mins) for SFC were considerably shorter than those required for HPLC (15 - 30 mins).

Stability of fungicides in aqueous buffer solutions. A necessary part of the above studies is confirmation that neither the artificial media used in some of the studies nor any of the work-up procedures affected the chiral integrity of the enantiomers. This was particularly so for the fungal metabolism work where a sterile liquid medium containing glucose and yeast extract was used, the pH value varying according to the fungus cultured. In both aqueous buffer solutions (pH values 4-8) and fungal cultures the enantiomers of

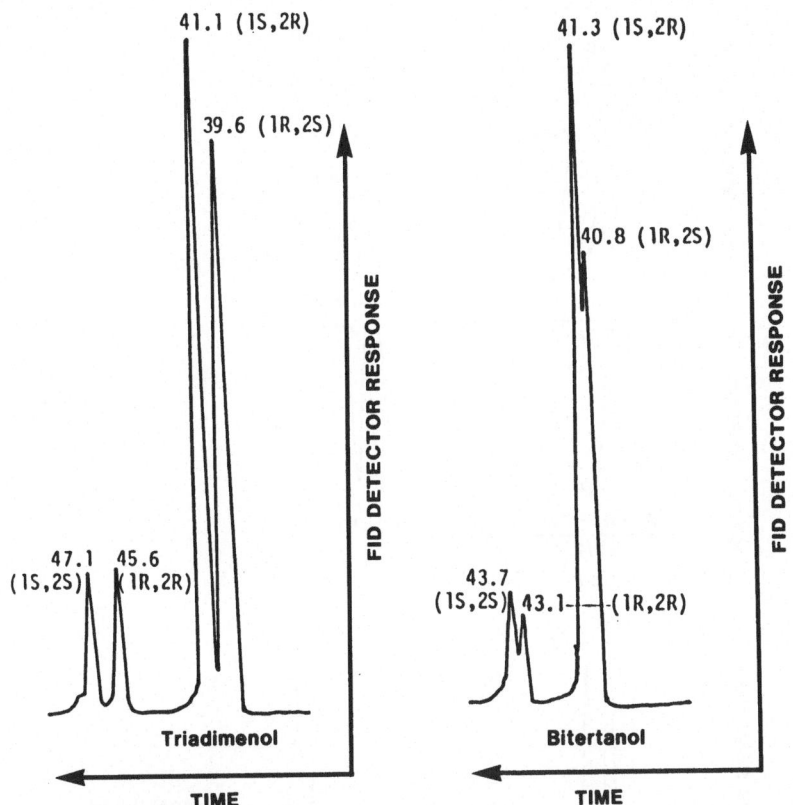

Fig. 2.   GC separation of triadimenol and bitertanol enantiomers (time in minutes) using "Chirasil Val". Enantiomer assignments were made by comparison of their retention times to those of authentic specimens.

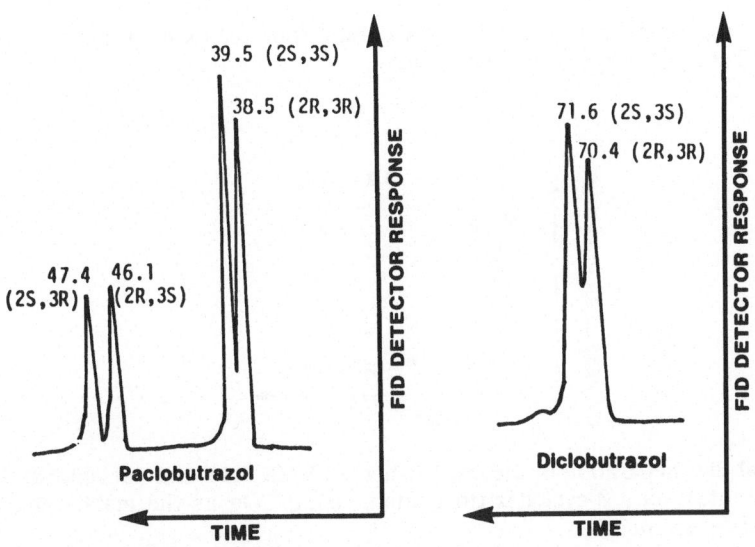

Fig. 3.   GC separation of paclobutrazol and diclobutrazol enantiomers (time in minutes) using "Chirasil Val". Enantiomer assignments were made by comparison of their retention times to those of authenitc specimens.

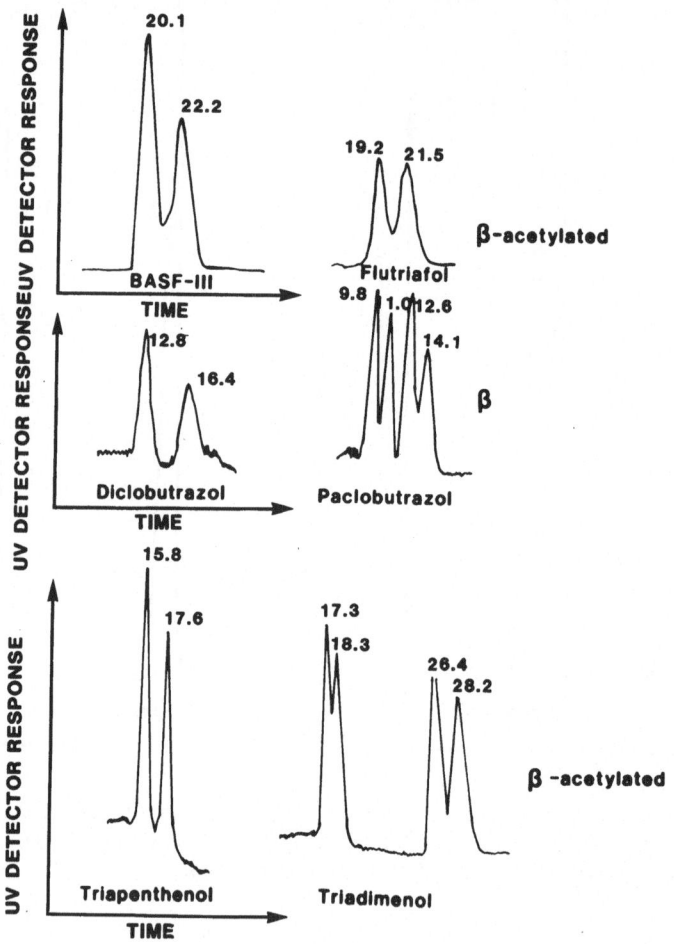

Fig. 4.   HPLC separation of enantiomers of six chiral traizoles (time in minutes) using β and β acetylated cyclodextrin columns.

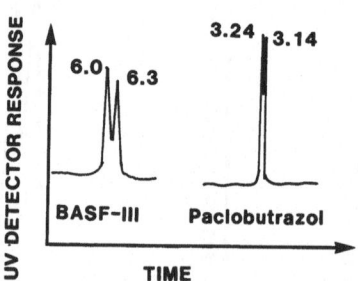

Fig. 5.   Chiral SFC separation of the enantiomers of BASF-111 and paclobutrazol (time in minutes) on a β-cyclodextrin column using CO_2 as the mobile phase and methanol as modifier.

triadimenol were stable.   However, the corresponding enantiomers of triadimefon were racemised at pH values greater than 5 with the rate of racemisation increasing with increasing pH. At pH 8 an equilibrium was reached within 16h.

Metabolism of triadimefon with fungi. Gastonyi et al,[13] have proposed that the activity of triadimefon results from its reduction by the target fungi to triadimenol. More recently[2,9,10] it has been suggested the activity may relate specifically to the production of the 1S, 2R enantiomer which is the most fungitoxic of the four triadimenol enantiomers (Fig.6). Thus, both the rate of reduction of triadimefon and the proportion of 1S, 2R-triadimenol formed are likely to be factors determining the sensitivity of fungi to triadimefon. We have investigated this selective reduction for 15 species of fungi[2] and four examples are discussed below.

Chiral GC indicated that different fungi produce different proportions of the four enantiomers of triadimenol. Chromatograms for *G. graminis, Botrytis cinerea, Fusarium culmorum* and *Cladosporium cucumerinum* are shown in Fig.7. and the quantitative results summarized in Table 1. The activity of triadimefon against a particular fungus appeared not to relate simply to the extent of its conversion to the 1S, 2R enantiomer of triadimenol. For example, in *Fusarium culmorum* although there was a substantial conversion of triadimefon to 1S,2R triadimenol the fungus was not very sensitive to triadimefon. This may have been due to poor penetration of 1S, 2R triadimenol to the site of action or poor inhibition of the target enzyme (sterol C-14 demethylase). There are probably three main criteria which determine the sensitivity of a fungus to applied triadimefon, i) its ability to convert triadimefon to triadimenol, ii) its ability to selectively produce the 1S, 2R enantiomer of triadimenol and iii) the intrinsic sensitivity of the fungus to the individual enantiomers.

For example, a fungus could convert triadimefon to triadimenol in low amounts and the proportion of the 1S,2R enantiomer could also be small but the fungus could still be extremely sensitive to the enantiomer. This is the case for *G.graminis* which shows a high conversion of triadimefon to triadimenol with the 1S,2R enantiomer representing only 1% of the triadimenol produced (Table 1).

Metabolism of traidimefon by different barley cultivars after leaf application. It is now accepted that the first step in the major metabolic pathway for triadimefon in plants[13] as in fungi[2], is reduction to triadimenol. It was decided to compare the rate and chirality of this reduction in three cultivars of spring barley (Golden Promise, Triumph and Patty). Chromatographic analysis of both the acetone leaf surface washings (Fig. 8) and leaf macerates at 7 and 21 days after treatment showed no differences in the triadimenol enantiomer distributions, only the 1S,2R and 1R, 2R forms being produced in significant amounts. There were no major differences in rates of reduction; in each cultivar little triadimefon remained after 7 days. Therefore we concluded that selective metabolism of triademefon is not an important factor in the observed differential control of powdery mildew on these barley cultivars by triadimefon.

Metabolism of triadimenol enantiomers by whole plant and soil after seed treatment. This study aimed to compare the metabolism and fate of the individual enantiomers of triadimenol in barley plants and soil after application of the compounds to seeds. Analysis of the radiolabelled residues in plants involved comparisons of the uptake of individual enantiomers, the degree and type of conjugation and their metabolism. A further important aspect of this work was the determination of the chiral stability of the enantiomers within plants since this information would prove invaluable when considering the suitability of the 1S,2R enantiomer for commercial development. The 1S,2R, 1R,2S enantiomers had comparable rates of uptake and conjugation in barley plants although the 1S,2R form was the only one to be conjugated through the secondary hydroxyl group. On the other hand, the uptake of the 1S,2S enantiomer was greater than the other enantiomers and it also showed significant differences in its metabolism and the type of conjugates formed[12].

Chiral analysis of purified triadimenol fractions on "Chirasil Val" proved impossible because, even after substantial clean-up of extracts, contaminants co-eluted with the triadimenol enantiomers. This problem was overcome by using six chiral Pirkle HPLC columns in series which were connected with a radio HPLC detector. The chiral

Table 1. Proportions of Triadimenol Enantiomers Formed and % of Residual Triadimefon After Metabolism of Traidimefon by Fungi Growing in Liquid Culture for 48 h.

	Triadimefen remaining (%)	Entiomeric composition of triadimenol			
		1R2S	1S2R	1R2R	1S2S
Gaeumannomyces graminis	9	1	1	43	55
Botrytis cinerea	49	76	6	0	18
Fusarium culmorum	0	0	66	34	0
Cladosporium cucumerinum	0	23	38	22	18

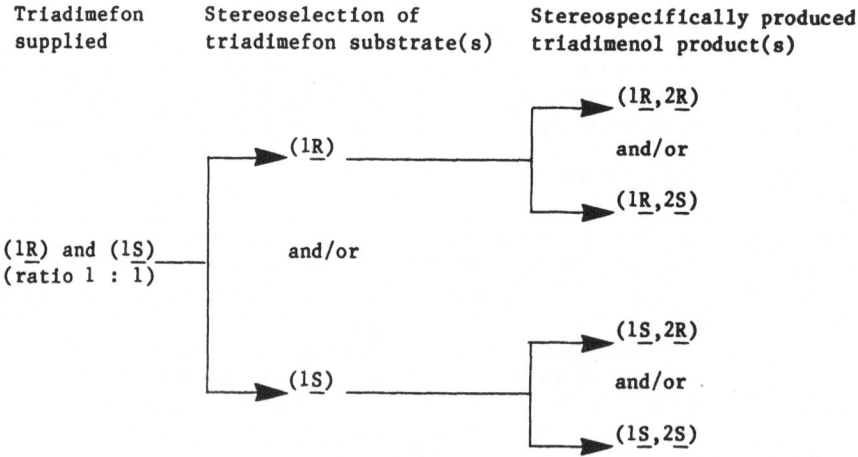

Fig. 6. Schematic of reduction of triadimefon to triadimenol.

chromatograms (Fig.9) indicated that the 1S,2R, 1R,2R and 1S,2S enantiomers were not racemised or epimerised within plants. However, the 1R,2S enantiomer although showing no evidence of racemisation, was epimerised to the 1R,2R form in a significant amount (23%). At this stage of the work it was not possible to determine whether the epimerisation had occurred in the plant or in the soil followed by uptake into the plants. Analysis of the soil extracts by TLC, prior to any clean-up showed that the 1R,2S enantiomer had been epimerised significantly more in soil (50%) than in plants and that the 1S,2R enantiomer had also been partly epimerised (25%). These findings needed confirmation by chiral analysis, but the Pirkle columns used for the plant analysis no longer gave sufficient resolution having apparently deteriorated, and an alternative was therefore sought. The β-acetylated cyclodextrin was capable of resolving the enantiomers of the 1RS,2SR diastereoisomer but not those of the 1RS,1RS form. However, when a soil extract was applied to the column the resolution of the first pair of enantiomers became poor and radio sensitivity decreased significantly. The samples were then analyzed by "Chirasil Val" with ECD after further clean-up by TLC using multiple developments. However, it was still not

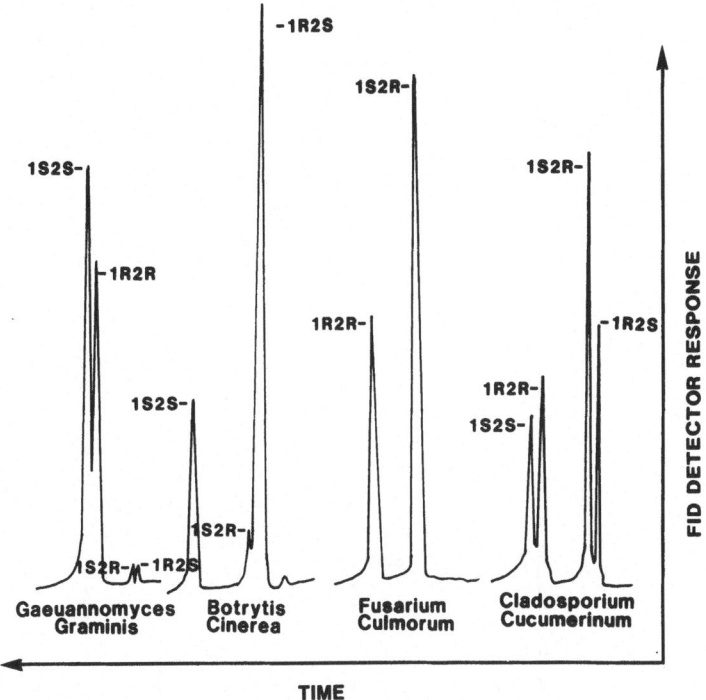

Fig. 7. Chiral GC ("Chirasil Val") of triadimenol produced during fungal metabolism of triademefon.

possible to assign any triadimenol peaks in the chromatogram. The soil extracts are still awaiting confirmation of their enantiomeric content.

Separation of Triazole Plant Growth Regulators

Both PGRs examined, BASF 111 and triapenthenol, were resolved by the β-acetylated cyclodextrin HPLC column (Fig. 4). Each product comprises two enantiomers (BASF 111 is predominantly one diastereoisomer) and by comparison with paclobutrazol and dichlopentezol it is probable that one enantiomer would be fungicidal, the other having plant growth regulatory activity. The latter could be produced by inhibition of gibberellin and/or sterol biosynthesis. In order to investigate the biological properties of these enantiomers mg quantities would be needed to be produced by preparative chiral techniques. The analytical column would then be used in the course of the studies of the above using the individual enantiomers.

CONCLUSIONS

The results reported above illustrate the increasing role of chiral analysis as a component of research on agrochemicals. Only by studying the separate enantiomers of a chiral pesticide can a complete understanding of its biological performance be obtained. However, it would prove too expensive for us to evaluate the available chiral columns for all our pesticides. Therefore it would be useful if the separation of compounds by chiral chromatography could be made more predictable. Also, for studies of of metabolism it would be most beneficial if the GC chiral columns could be successfully coupled to a radio-detector. Such improvements would make investigation of the performance of agrochemicals at the enantiomer level considerably more efficient.

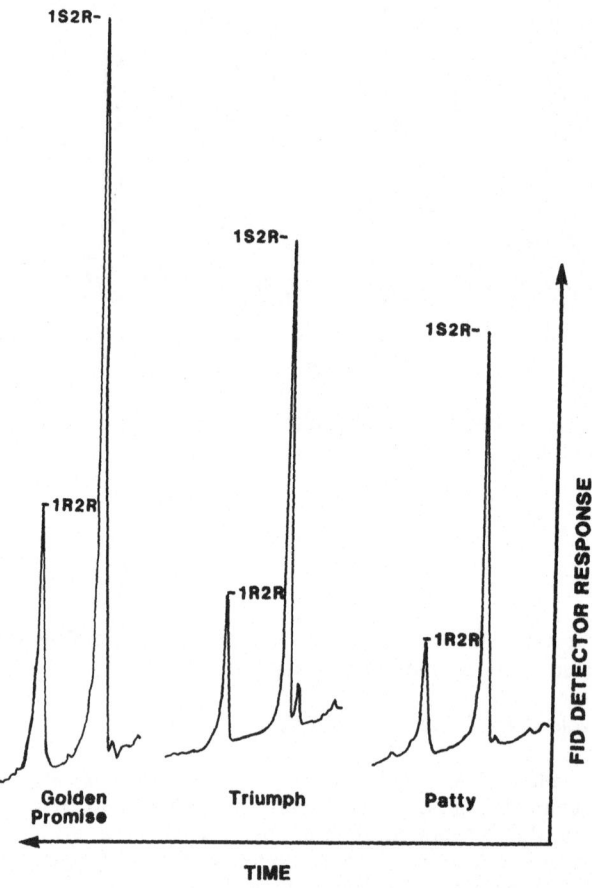

Fig. 8.   Chiral GC ("Chirasil Val") of the acetone leaf surface washings containing triadimenol produced by the metabolism of triadimefon in the leaves of three different barley cultivars following foliar application of the fungicide.

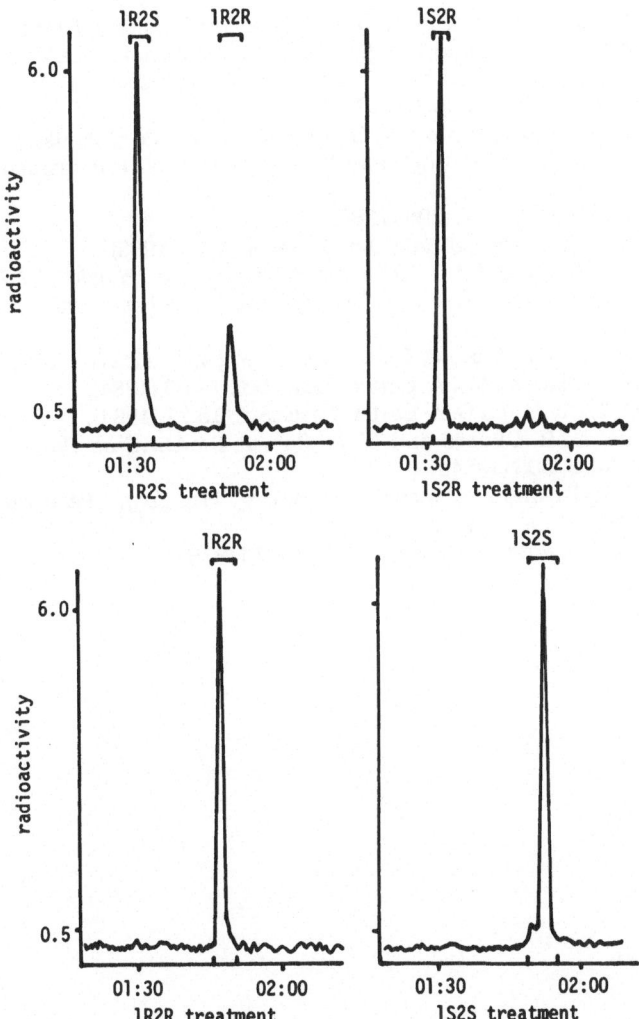

Fig. 9.   Chiral HPLC chromatograms (Pirkle columns) of triadimenol isolated from barley plants after seed treatment with individual enantiomers.

Acknowledgements

Long Ashton Research Station is financed by the Agricultural and Food Research Council. The authors express their gratitude to Bayer AG for providing radiolabelled material and to Dr D. Games, University College Cardiff, for the provision of SFC facilities. They also thank Drs N.H. Anderson and R.S. Burden for useful discussions and advice and Dr D.M. Johns of Technicol, Stockport, for the supply of HPLC columns.

REFERENCES

1. W. Kramer, K. H. Buchel and W. Draber, "Pesticide Chemistry, Human Welfare and the Environment," P. Doyle and T. Fujita, eds, Vol. 1, Pergamon, Oxford, p. 223 (1983).
2. A. H. B. Deas, G. A. Carter, T. Clark, D. R. Clifford and C. S. James, *Pestic.Biochem.Physiol.*, 26:10 (1986).
3. Y. Funaki, Y. Ishiguri, T. Kato and S. Tanaka, "Pesticide Chemistry, Human Welfare and the Environment," P. Doyle and T. Fujita, eds., Vol. 1, Pergamon, Oxford, p. 309, (1983).
4. B. Sugavanum, *Pestic.Sci.*, 15:296 (1984).
5. P. Hedden and J. E. Graebe, *J.Plant Growth Regul.*, 4:11 (1985).
6. R. S. Burden, T. Clark and P. J. Holloway, *Pestic.Biochem. Physiol.*, 27:289 (1987).
7. A. H. B. Deas and G. C. Carter, *Proc.Br.Crop Prot.Conf. - Pests and Diseases*, Vol. 1:835 (1986).
8. A. H. B. Deas and D. R. Clifford, *Pestic.Biochem.Physiol.*, 22:276 (1984).
9. A. H. B. Deas, T. Clark and G. A. Carter, *Pestic.Sci.*, 15:63 (1984).
10. A. H. B. Deas, T. Clark and G. A. Carter, *Pestic.Sci.*, 15:71 (1984).
11. A. H. B. Deas, D. R. Clifford and G. A. Carter, *Proc.Br.Crop Prot.Conf. - Pests and Diseases*, Vol. 3, 905 (1984).
12. T. Clark, K. Vogeler and I. Ishikawa, *Proc.Br.Crop Prot.Conf. - Pests and Diseases*, Vol. 1, 475 (1986).
13. M. Gastonyi and G. Josepovits, *Pestic.Sci.*, 10:57 (1979).

RECENT DEVELOPMENTS IN ENANTIOMER SEPARATION BY

COMPLEXATION GAS CHROMATOGRAPHY

Volker Schurig* and Rainer Link

Institut für Organische Chemie der Universität
Auf der Morgenstelle 18, D-7400 Tübingen, FRG

SUMMARY

The thermodynamics of enantiomer separation by complexation gas chromatography is derived and verified by experiments. The influence of structural changes at the C(10) position of camphor upon enantiomer discrimination by camphorato-Chirametal stationary phases is probed. The preparation of a novel Chirasil-Metal stationary phase for enantiomer separation in complexation gas chromatography is described.

INTRODUCTION

The unambiguous determination of enantiomeric compositions and absolute configurations is an important analytical task in research concerned with the synthesis, characterization and use of chiral compounds. The "enantiomeric excess" (ee) provides a quantitative criterion to describe the success of an enantioselective process. The precise determination of enantiomeric composition may also be warranted in the characterization of natural products such as flavors, fragrances or pheromones, in the monitoring of the enantiospecificity of enzymic reactions, in the detection of racemization in "chiral pool" synthesis, in peptide synthesis/hydrolysis as well as in the study of reaction mechanisms. The availability of reliable and precise techniques for the correct determination of ee is therefore of great importance in order to cope with the analytical requirements of contemporary stereochemistry.

The quantitative separation of enantiomers by gas chromatography on chiral, non-racemic, stationary phases[1-5] constitutes a powerful tool for enantiomer analysis because of speed, simplicity, reproducibility and sensitivity[6].

In complexation gas chromatography[7] an electronically and coordinatively unsaturated transition metal compound, capable of exerting chemical affinity toward solutes with suitable chemical functionalities, is added to the stationary liquid phase. Due to the fast and reversible chemical interaction between the additive and the solute, the separation of structural isomers and isotopomers can be carried out by virtue of chemical selectivity. The extension of chemical selectivity into enantioselectivity permits also the discrimination of optical isomers on chiral, non-racemic metal coordination compounds employed as additives to the stationary phase[8,9]. Thus, volatile racemic ethers, ketones, alcohols, acetals, esters as well as racemates of other classes of compounds have been quantitatively separated into enantiomers on manganese(II)-, cobalt(II)- or nickel(II)-bis[3-(heptafluorobutanoyl)-(1R)-camphorate] (1a) and related Chirametal stationary phases[10-12]. In contrast to complementary methods of gas chromatographic enantiomer separation[2,3], solute derivatization is generally not required by complexation gas chromatography, rendering enantiomer analysis by this methodology very straightforward indeed.

The use of complexation gas chromatography in enantiomer analysis has previously been summarized[5,6], and many novel applications in various fields of (bio)chemical sciences have been reported using glass or fused silica high-resolution open-tubular columns coated with Chirametal stationary phases[13,14] from our group[15-19] and from other laboratories[20-25].

At present, the thermal stability and the enantioselectivity of Chirametal stationary phases is limited. Thus there is no single Chirametal stationary phase available which will perform the enantiomer separation of most volatile racemates. An improvement of enantioselectivity may be brought about *inter alia* by structural changes of existing Chirametal stationary phases such as nickel(II)*bis*[3-(heptafluorobutanoyl)-(1R)-camphorate] (1a).

$$R = \quad -CH_3 \qquad \textbf{1a}$$
$$-CH=CH_2 \quad \textbf{1b}$$
$$-CH_2-CH_3 \quad \textbf{1c}$$

Here we report on chemical modifications of the camphor moiety at the C(10) position of camphor.

Another important objective is the preparation of thermostable Chirametal stationary phases which are inert toward decomposition and bleeding-off the column at high separation temperatures. At present, the temperature limit of the standard Chirametal stationary phase nickel(II)*bis*[3-(heptafluorobutanoyl)-(1R)-camphorate] (1a) is approximately 120°C. Above this temperature, a decrease of interaction with coordinating solutes gradually commences. One obvious approach to the lowering of the bleeding rate of Chirametal stationary phases is the coupling of the metal-β-diketonate to a polysiloxane matrix analogous to the well established preparation of thermally stable "Chirasil-Val"[2] stationary phases. Preliminary results on the preparation of a "Chirasil-Metal" stationary phase are described here.

It should be recognized, however, that elevated separation temperatures are *per se* detrimental to gas chromatographic enantiomer separation as a result of the second law of thermodynamics governing enantioselectivity. An "iso-enantioselective temperature" T_{iso} has been suggested as important in enantiomer separation on chiral stationary phases: at T_{iso} separation of enantiomers will vanish while inversion of the order of elution is predicted above this temperature[26,27]. In this report the derivation of thermodynamic parameters governing chiral recognition based on relative retention data is advanced and the extrapolation to T_{iso} is (tentatively) carried out in a study of the temperature dependence of enantiomer separation by complexation gas chromatography.

Hitherto, enantiomer separation by complexation gas chromatography has been investigated by a rather empirical approach[5-7]. The present study on thermodynamic aspects of chromatographic enantiomer separation is aimed at elucidating inherent theoretical requirements for efficient chiral recognition.

EXPERIMENTAL

Instrumentation

Carlo-Erba gas chromatographs, Fractovap 2101 and 2350, equipped with FID and suitable for operation with glass open-tubular columns, were used. A Carlo-Erba gas chromatograph, HRGC 5300 MEGA, was employed for fused silica open-tubular columns. The carrier gas was nitrogen. *Caution:* hydrogen must not be employed as carrier gas in complexation gas chromatography.

Splitting devices were set 1:100. In order to avoid overloading, which results in peak-tailing and peak-broadening, the instrument was set at its highest sensitivity at a tolerable signal to noise ratio.

Open-tubular Columns

The preparation of glass and fused silica open-tubular columns for use in complexation gas chromatography has been described previously in detail[13,14,28]: Duran glass tubing (obtained from Schott Ruhrglas, Mainz, FRG) was drawn to capillaries of 0.25 mm I.D. using a Hupe & Busch glass-drawing machine.

Acid leaching and rinsing: 90% of the capillary column was filled with aqueous 6 N hydrochloric acid, sealed under vacuum and heated to 150°C for 24 h. The ends were opened and the capillary column was rinsed with approximately three volumes of 0.01 N hydrochloric acid and 1 ml of methanol. Finally, the column was dried for 2 h at 250°C under a stream of dry nitrogen.

Deactivation with DPTMDS (diphenyltetramethyldisilazane): The column was dynamically coated with a solution of DPTMDS in n-pentane (1:1, v/v). The column was sealed under vacuum and heated up to 200°C. The temperature was increased to 390°C at a rate of 3°C/min and held there for 12 h. The column ends were opened and the capillary column was rinsed with 1 ml of n-pentane, 3 ml of methanol and 2 ml of diethyl ether[29].

Coating: The column was coated with the required amount of the Chirametal stationary phase in polysiloxane dissolved in dichloromethane/n-pentane (1:9, v/v)(0.2-0.5%) by the static method.

For static coating of the polymeric phase Chirasil-Metal 3 0.5-0.8% solutions in n-pentane/CH_2Cl_2 (9:1) were used.

After coating the columns were conditioned by temperature programming from 60°C to 90°C at 1°C/min and 3 h holding at 90°C under nitrogen-gas purge (190°C for the polymeric phase Chirasil-Metal).

Reference Solutes

For thermodynamic measurements (cf. also[11]) methane was coinjected to determine the dead volume (gas hold-up) of the open-tubular columns. n-Octane (and in the case of 2 n-dodecane) was also coinjected as non-coordinating reference standard for the determination of r.

Retention times have been obtained graphically and via computer-aided data acqui-sition (Shimadzu C-R3A). Excellent agreement was observed.

Organic Solutes

Most of the solutes were obtained from commercial sources (Fluka, Buchs, Switzerland; Merck, Darmstadt, GFR; Aldrich-Chemie, Steinheim, GFR). Oxiranes not available com-mercially were prepared from alkenes by epoxidation with m-chloroperbenzoic acid[12]. Oxiranes were handled with appropriate care in closed systems. Chalcogram 2 was prepared according to[30].

Preparation of the β-diketonate Ligands

(3-Heptafluorobutanoyl)-(1R)-camphor was prepared according to the general procedure of Whitesides et al. by acylation of (1R)-(+)-camphor with heptafluorobutanoylchloride[31]. (3-Heptafluorobutanoyl)-(1S)-10-methylene-camphor has been obtained by acylation of (1S)-10-methylenecamphor[32] and heptafluorobutanoylchloride as described for (3-heptafluorobutanoyl)camphor.

b.p.: 55-60°C/0,03 Torr;

α_D^{25} + 14,4° (0,1 dm; neat);

^{13}C - NMR (CDCl$_3$)

δ 211.9, 149.0, 130.3, 119.9, 63.9, 50.6, 47.6, 26.6, 20.5, 18.3;

MA (m/e) 361 (18%), 360 (M$^+$, 100%), 317 (21%), 191 (23%), 163 (28%), 135 (20%), 107 (24%), 91 (24%), 79 (20%), 53 (24%)..

(3-Heptafluorobutanoyl)-(1R)-10-methyl-camphor was prepared from (3-heptafluorobutanoyl)-(1S)-10-methylene-camphor by catalytic hydrogenation. Thus, 2.5 g (3-heptafluorobutanoyl)-(1S)-10-methylene-camphor were dissolved in methanol. After saturation of the solution with nitrogen, 250 mg of palladium (10% on charcoal) were added. Hydrogenation was continued until the reduction ceased. The solution was then filtered and concentrated in vacuo. Distillation of the remaining oil at 58-60°C (0.03 mm Hg) yielded 0.62 g of a colorless liquid.

α_D^{20} + 15,6° (0,1 dm; neat);

^{13}C-NMR (CDCl$_3$) δ 214.5, 148.5, 120.7, 60.9, 49.5, 47.8, 26.6, 26.4, 20.8, 18.8, 17.9, 9.4;

MS (m/e) 363 (18%), 362 (M$^+$, 100%), 333 (76%), 291 (32%), 165 (34%), 137 (20%), 95 (24%), 55 (45%), 29 (22%).

Metal Chelates

All metal chelates were prepared from the β-diketonates as described previously[12]. Strict adherence to the published procedures is strongly recommended. Chirametal stationary phases prepared and purified by alternate routes may exert different physical and chomatographic properties.

Nickel(II)*bis*[3-heptafluorobutanoyl)-(1S)-10-methylene-camphorate] 1b: green powder, m.p.: 105-110°C.

$[\alpha]_D^{20}$ + 101,0° (c=0,1; CHCl$_3$);

MS (m/e) 776 (M$^+$,^{58}Ni, 100%), 748 (80%), 418 (84%), 359 (78%), 95 (84%), 55 (82%);
IR (KBr, cm^{-1}) 3400, 2960, 1640, 1520, 1340, 1220, 1110, 905, 785, 745, 725.
Nickel(II)*bis*[3-heptafluorobutanoyl)-(1R)-10-methylene-camphorate] 1c: green powder, m.p.: 130-140°C.

$[\alpha]_D^{25}$ + 138.7° (c=0.3; CHCl$_3$);

MS (m/e) 780 (M$^+$,^{58}Ni, 82%); 752 (54%), 417 (38%), 362 (22%), 166 (18%), 122 (38%), 109 (38%), 95 (100%), 83 (28%), 69 (38%), 67 (30%);
IR (KBr, cm^{-1}) 3420, 2960, 1710, 1650, 1530, 1340, 1230, 1110, 900, 745.

Synthesis of the Chiral Polysiloxane Stationary Phase Chirasil-Metal 3

(3-Heptafluorobutanoyl-)(1S)-10-(dimethoxymethylsilyl)-methylcamphor was prepared by hydrosilylation of (3-heptafluorobutanoyl)-(1S)-10-methylene-camphor with methyldimethoxysilane. Under an atmosphere of nitrogen, 0.1 ml newly prepared catalyst (H$_2$PtCl$_6$x6H$_2$O, 1% in 2-propanol) were added to 1.5 g (4.1 mmol) of (3-heptafluorobutanoyl)-(1S)-10-methylene-camphor. After heating at 50°C the solution was treated with 0.5 ml (4.1 mmol) methyldimethoxysilane. After stirring for 12 hours at 50°C, approximately 0.1 ml methyldimethoxysilane were added until the reaction was complete. Methanolysis of the silylated β-diketonate carbonyl groups was performed by adding 10 ml methanol and refluxing for several hours. After methanol was removed in vacuo, the residue was purified by column chromatography (silica, toluene/high boiling petroleum ether 3:1 v/v). Distillation of the red-brown residue at 40-45°C (0.03 mm Hg) yielded 0.5g (24.4%) of a colorless liquid. $[\alpha]_D^{25}$ + 11,1° (0,1 dm; neat);

^{1}H-NMR (CDCl$_3$,ppm) δ 0.12 (s,3H), 0.75-1.94(m), 2.72-2.84 (m,1H), 3.51(s,6H).

Trimethylsilanol was prepared from hexamethyldisilazane according to the literature[33]. Hydrolysis of diethoxymethylvinylsilane was carried out by mixing of 8g (0.05 mol) diethoxymethylvinylsilane, 5g ethanol and 20g H$_2$O and stirring for 1h at 90°C. Afterwards the product was extracted with ether and dried by azeotropic distillation with benzene.

IR (cm^{-1}) 3050, 3023, 2960, 1600, 1410, 1260, 1090, 1020, 960, 820, 790.
GC-MS investigations proved the predominance of three-, four- and five-membered cyclic siloxanes in the product.

For polymerization, nickel(II)bis[(3-heptafluorobutanoyl-)(1S)-10-(dimethoxy-methylsilyl)-methylcamphorate] has been prepared acc. to [12] and the crude product (0.5 g) was immediately hydrolyzed by adding 7.5 ml methanol and 40 ml H$_2$O. After stirring for 10 hours at room temperature, the product was extracted with dichloromethane and, after evaporation of the solvent, dissolved in methanol. The filtered solution was concentrated in vacuo and the residue (0.45g) was treated with 2.25g of hydrolyzed diethoxy-methylvinylsilane and 0.05g trimethylsilanol. After evaporation of the solvent, polymerization was accomplished by adding 7 µl of the tetramethylammonium hydroxide (TMAH) solution in methanol (20% w/v) and heating at 110-120°C for four hours. Again 7 µl of the tetramethylammonium hydroxide solution were added and stirring was continued at 110-120°C for another ten hours. Due to an interaction of the catalyst and the metal chelate, more catalyst has to be added thereafter in small portions until the viscosity of the polymer ceased to change (catalyst coordinated to the metal chelate is removed during conditioning of the columns at 190°C). In order to remove low molecular-weight material, the polymer was dissolved in dichloromethane and fractionated three times by addition of an equal volume of methanol. The resulting polymer was exhaustively washed with methanol and afterwards end-capped at room temperature using 1,3-divinyltetramethyldisilazane in dichloromethane. After evaporation of the solvent, the polymer was stored in bottles under nitrogen.

RESULTS AND DISCUSSION

1) *Thermodynamics of Enantiomer Separation by Complexation Gas Chromatography*[33]

Thermodynamic parameters describing chiral recognition are readily accessible from relative retention data of enantiomers separated on Chirametal stationary phases by complexation gas chromatography[12].

Enantiomer separation by complexation gas chromatography requires (i) that the stabilities of the diastereomeric donor-acceptor association complexes are different, (ii) that complex formation is reversible, and (iii) that the equilibrium is established rapidly.

While the chemical interaction between the selectand and the selector (selectand refers in the present context to a racemic solute which is resolved on the non-racemic Chirametal stationary phase, referred to as the selector) is expressed by the association constant K, or the free enthalpy of association -ΔG°, respectively, the difference of -ΔG° for a pair of enantiomers, $-\Delta_{R,S}(\Delta G°)$ represents a thermodynamic quantity for enantiomer discrimination. (R and S arbitrarily denote oppositely configured enantiomers whereby the S-enantiomer precedes the R-enantiomer on elution from A, i.e., K$_R$ > K$_S$.)

Thermodynamic data of the selectand-selector association equilibrium can be obtained from relative retention data by complexation gas chromatography as follows: When a selectand B is gas chromatographed on a stationary phase containing the solution of a selector A (e.g. a Chirametal selector) in an inert solvent S (e.g. a polysiloxane or squalane) the retention of B is not only dependent on the physical partition equilibrium between the gaseous and liquid phases but is also determined by the reversible and rapid chemical association equilibrium in the liquid phase.

$$A + B = AB,$$

The chemical interaction leads to a retention-increase R' which is related to the association constant K and to the activity of A in S (a_A) by the linear relationship[35].

$$K \cdot a_A = \frac{r - r_0}{r_0} = R' \tag{1}$$

R' can be calculated from the readily accessible relative retention data r_0 and r, where r = relative adjusted (= corrected for the dead-volume of the column) retention of B with respect to an inert reference standard, not interacting with A, on a column containing the activity a_A in S, and r_0 = relative adjusted retention of B with respect to the inert reference standard on a reference column containing the pure solvent S.

With

$$K = \frac{a_{AB}}{a_A \cdot a_B} \tag{2}$$

it follows from equation (1) that the retention-increase R' defines the activity fraction of the complexed vs. uncomplexed selectand B in the liquid phase (A in S), i.e.,

$$R' = \frac{a_{AB}}{a_B} \tag{3}$$

Because only a trace of the selectand B (approximately 10^{-8} g) as well as a dilute solution of the selector A in the solvent S (0.05-0.1 m) is employed in complexation gas chromatography, equations (1) and (3) can be simplified to

$$K_{(m)} \cdot m_A = \frac{r - r_0}{r_0} = \frac{m_{AB}}{m_B} = R' \tag{4}$$

The molality concentration scale is chosen because m is temperature independent and the weight rather than the volume of S is determined for practical purposes. K and $-\Delta G°$, and, by measurements at different temperatures, the corresponding Gibbs-Helmholtz parameters $-\Delta H°$ and $\Delta S°$, can be obtained from equation (4). When R' is distinct for a pair of enantiomers on the Chirametal selector A, peak separation will occur according to equations (5) and (6)

$$\frac{R'_R}{R'_S} = \frac{K_R}{K_S} = \frac{r_R - r_0}{r_S - r_0} \tag{5}$$

or

$$-\Delta_{R,S}(\Delta G°) = RT \cdot \ln \frac{R'_R}{R'_S} \tag{6}$$

(note that eq. (6) is independent of the actual concentration of the Chirametal selector in the solvent S, m_A, and that r_0 is alike for enantiomers. Thus, possible errors in $-\Delta_{R,S}(\Delta G°)$ are greatly reduced and the quantity should thus be highly precise).

The thermodynamic quantities for enantiomer discrimination $-\Delta_{R,S}(\Delta G°)$, and, by measurements at different temperatures, the corresponding Gibbs-Helmholtz parameters

96

$\Delta_{R,S}(\Delta H°)$ and $\Delta_{R,S}(\Delta S°)$ are accessible from the difference of the retention-increases R' of the enantiomers:

$$RT \cdot \ln \frac{K_R}{K_S} = RT \cdot \ln \frac{R'_R}{R'_S} = -\Delta_{R,S}(\Delta G°) = -\Delta_{R,S}(\Delta H°) + T\Delta_{R,S}(\Delta S°) \qquad (7)$$

The validity of equation (4) as well as the high precision of $-\Delta_{R,S}(\Delta G°)$ have previously been verified in an investigation of four racemic alkyl-substituted oriranes at five concentrations of manganese(II)bis[3-(heptafluorobutanoyl)-(1R)-camphorate] in squalance ($m = 0.05$-0.15)[10]. Here we report on another verification of equation (4). Thus, according to eq. (8) r_0 of the selectand can be extrapolated from the relative retentions r^R and r^S obtained at two concentrations (1 and 2) m of the selector:

$$r_0 = \frac{r_1^R r_2^S - r_1^S r_2^R}{(r_1^R + r_2^S) - (r_1^S + r_2^R)} \qquad (8)$$

The following relative retentions r (with respect to n-dodecane at 93°C) have been measured for the enantiomers of E-2-ethyl-1,6-dioxaspiro[4,4]nonane, "chalcogran" E-2 (the principal aggregation pheromone of the Kupferstecher, Pityogenes chalcographus (L.)[36] (formula 2).

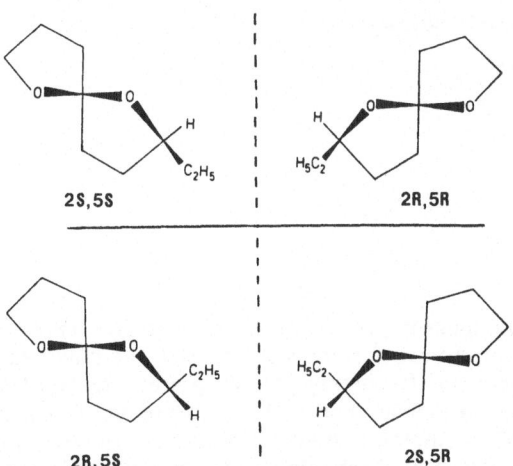

2S,5S   2R,5R

2R,5S   2S,5R

on a 25 m x 0.25 mm (I.D.) fused silica capillary column coated with two (arbitrary) concentrations 1 and 2 of nickel(II)bis[3-heptafluorobutanoyl)-(1R)-camphorate] (1a) in SE 54 (polysiloxane):

$r_1^R = 0.95$ / $r_1^S = 1.13$
$r_2^R = 1.90$ / $r_2^S = 2.44$

r_0 is calculated to 0.475. Measurements of r_0 for E-2 between 80°C and 120°C at intervals of 5°C revealed a linear relationship between r_0 and T, furnishing an extrapolated value of 0.476 ± 0.001 for ro at 93°C which is in excellent agreement with the calculated figure. Comprehensive thermodynamic data of the selectand-selector association equilibrium between the four configuration isomers of chalcogran (Z- and E-2-ethyl-1,6-dioxaspiro[4,4]nonane) 2 on nickel(II)bis[3-heptafluorobutanoyl)-(1R)-camphorate] 1a in SE-54 have been determined in the following way (cf. Table 1).

First r_0 of Z- and E-chalcogran 2 (with respect to the inert reference standard n-dodecane) was measured between 80°C and 120°C with intervals of 5°C on a 25m x 0.25 mm fused silica open-tubular column containing only SE 54 (phenylvinylpolysiloxane).

Afterwards the same column was re-coated with ≈ 0.1 m nickel(II)bis[3-(hepta-fluorobutanoyl)-(1R)-camphorate] 1a in SE 54 and r for each of the Z- and E-enantiomeric pairs of chalcogran 2 (with respect to the inert reference standard n-dodecane) was measured at the same temperatures, cf. Fig. 1 and Table 1.

Analysis of the data revealed that the individual plot of R ln(R') against 1/T (one individual enantiomer) was not linear at all, probably because of the incertitude of m_A at different temperatures (e.g. due to changes of m_A by concentration gradients in the column, partial insolubility, decomposition or another unrecognized causes).

However, the plot for the respective racemic mixture (two enantiomers):

$$R \cdot \ln \frac{R'_R}{R'_S} = -\Delta_{R,S}(\Delta H°)/T + \Delta_{R,S}(\Delta S°) \tag{9}$$

was linear for the Z-enantiomers of 2 between 80°C and 105°C and for the (strongly coordinating) E-enantiomers of 2 between 80-120°C, cf. Scheme 1.

The following Gibbs-Helmholtz parameters were calculated from Table 1.:

Z-chalcogran 2:

$-\Delta_{R,S}(\Delta G°) = 0.25$ kcal.mole^{-1} (353.16K)
$-\Delta_{R,S}(\Delta H°) = 0.92$ kcal.mole^{-1}
$-\Delta_{R,S}(\Delta S°) = 1.90$ e.u.

E-chalcogran 2:

$-\Delta_{R,S}(\Delta G°) = 0.20$ kcal.mole^{-1} (353.16K)
$-\Delta_{R,S}(\Delta H°) = 0.81$ kcal.mole^{-1}
$-\Delta_{R,S}(\Delta S°) = 1.75$ e.u.

These data merit the following comments: Although the retention-increase R' (being proportional to the equilibrium constant of coordination) was only nearly half for the Z- as compared to the E-enantiomers, the thermodynamic quantities, $-\Delta_{R,S}(\Delta G°)$, $-\Delta_{R,S}(\Delta H°)$ and $-\Delta_{R,S}(\Delta S°)$ showed comparable values, that is, the degree of chiral recognition is found to be quite independent from the complexation strength of the selectand and the selector! This result is corroborated by empirical observations that separation factors of enantiomers may be quite independent from their degree of total chemical interaction with Chirametal stationary phases.

Moreover, the signs of $\Delta_{R,S}(\Delta H°)$ and $\Delta_{R,S}(\Delta S°)$ are such that they oppose each other in determining the sign of $\Delta_{R,S}(\Delta G°)$, that is, at low temperatures enantiomer separation is enthalpy-controlled while at high temperatures enantiomer separation is entropy-controlled. Indeed, the enantiomer which undergoes a stronger coordination forms an association complex (possessing a more negative enthalpy change) which is higher ordered (possessing a more negative entropy change). Dependent on the absolute values for $-\Delta_{R,S}(\Delta H°)$ and $-\Delta_{R,S}(\Delta S°)$ peak-coalescence (of the third kind) ($\Delta_{R,S}(\Delta G°) = 0$, i.e. no enantiomer separation) will commence at the so-called iso-enantioselective temperature T_{iso}[4,26,27]. For peak-coalescence phenomena of the first and second kind cf.[12].

It has been predicted[4,26,27], (and recently) verified experimentally, that above the iso-enantioselective temperature, T_{iso}, the elution order will be reversed (peak-inversion) and separation factors will increase again at raising separation temperatures with the (odd) consequence that the enantiomer undergoing stronger chemical bonding will be eluted first due to the importance of the entropic disorder of the association complex, see scheme 2.

Table 1. Retention data of enantiomer separation of chalcogran (Z and E-2-ethyl-1,6-dioxaspiro[4.4]nonane) 2 between 80-120°C measured on 25 m × 0.25 mm I.D. fused silica capillary column coated with ≈ 0.1 m nickel(II)bis[3-(heptafluorobutanoyl)-(1R)-camphorate] (1a) in SE-54.

	80°C	85°C	90°C	95°C	100°C	105°C	110°C	115°C	120°C
Z (r_0)	0.428	0.442	0.456	0.469	0.483	0.496	0.510	0.523	0.537
E (r_0)	0.443	0.456	0.469	0.482	0.495	0.508	0.521	0.534	0.547
Z(A) (r)	0.831	0.840	0.846	0.849	0.850	0.852	0.854	0.856	0.858
Z(B) (r)	0.964	0.961	0.956	0.949	0.940	0.933	0.925	0.919	0.911
E(C) (r)	1.648	1.631	1.615	1.595	1.565	1.535	1.510	1.479	1.450
E(D) (r)	2.153	2.069	2.037	1.976	1.914	1.850	1.790	1.730	1.670
Rln(R_B'/R_A')	0.5597	0.5271	0.4929	0.4641	0.4357	0.4073	0.3724	0.3442	0.3035
Rln(R_D'/R_C')	0.7007	0.6621	0.6226	0.5865	0.5606	0.5313	0.4951	0.4677	0.4329
$10^3/T$	2.8316	2.7920	2.7536	2.7162	2.6798	2.6444	2.6099	2.5763	2.5435

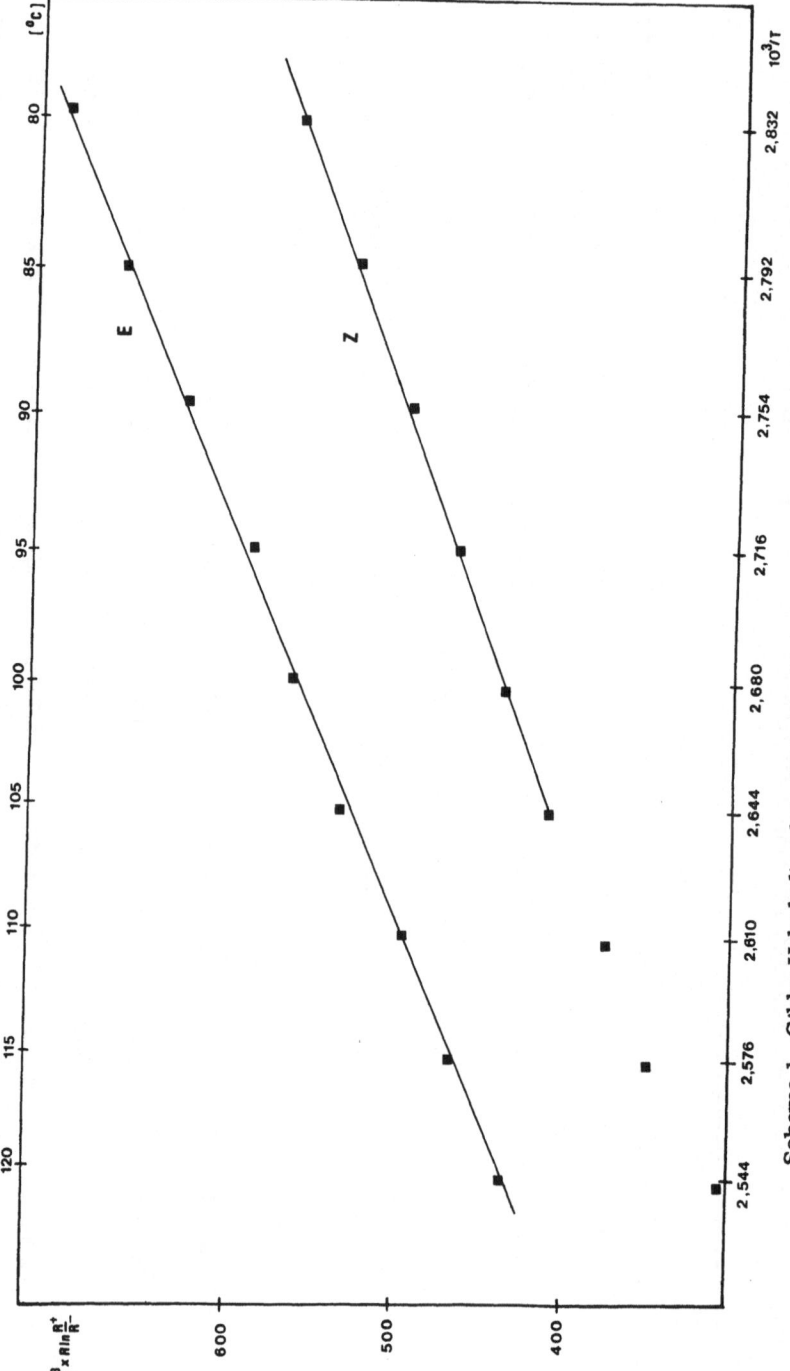

Scheme 1. Gibbs-Helmholtz-plot of $R \ln(R_R/R_S)$ versus $1/T$ for Z- and E-chalcogran 2.

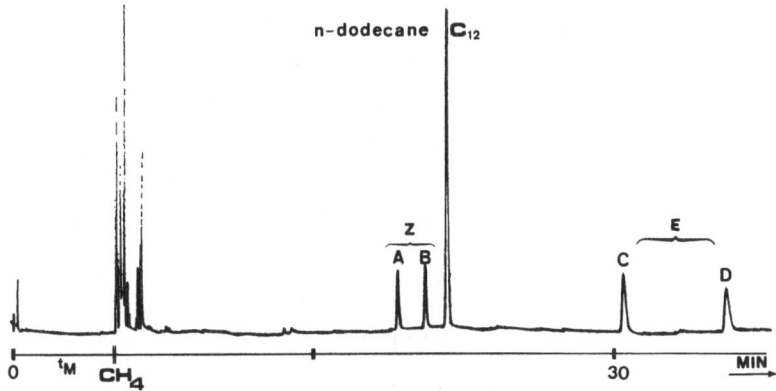

Fig. 1. Determination of thermodynamic data from relative retention data. Enantiomer separation of chalcogran (Z- and E-2-ethyl-1.6-dioxaspiro[4.4]nonane) (2) on a 25m x 0.25 mm I.D. fused silica capillary column coated with ≈ 0.1m nickel(II)bis[3-(heptafluorobutanoyl)-(1R)-camphorate] (1a) in SE-54.

Because of the (expected) temperature-dependence of $\Delta_{R,S}(\Delta H°)$ and $\Delta_{R,S}(\Delta S°)$ the estimation of T_{iso} based on the results of Table 1 is very risky indeed. We nevertheless undertake to report here extrapolated values for T_{iso} as calculated from eq. (9), cf. Scheme 2:

E-chalcogran 2: $T_{iso} \approx 484$ K (210°C)

Z-chalcogran 2: $T_{iso} \approx 463$ K (190°C)

Thus, peak-coalescence (no enantiomer separation) of Z- and E-chalcogran (2) on nickel(II)bis[3-(heptafluorobutanoyl)-(1R)-camphorate] (1a) is to be expected at approximately 200°C. Unfortunately, the thermal instability of the Chirametal stationary phase at this temperature prevents the verification of this prediction. It should be noted that at 150°C quantitative enantiomer separation still commences as the result of the high enantioselectivity in this particular selectand-selector system.

The foregoing considerations imply that temperatures below T_{iso} will always be favorable for enantiomer separation. Unfortunately, gas chromatographic enantiomer separation (in contrast to liquid and super-critical-fluid chromatography) has frequently to rely on high separation temperatures because of the involatility of many solutes. However, as predicted by the Gibbs-Helmholtz equation, enantiomer separation becomes independent from the separation temperature when $\Delta_{R,S}(\Delta H°)$ is zero, that is, when no chemical bonding between the selectand and selector takes place. Such a situation may arise, in principle, by chiral recognition due to physical inclusion.

2) Influence of Structural Changes at the C(10) Position of Camphor upon Enantioselectivity of Camphorato-Chirametal Stationary Phases[37]

Nickel(II)bis[3-(heptafluorobutanoyl)-(1R)-camphorate] (1a) represents a versatile Chira-metal stationary phase for enantiomer separation of racemic compounds belonging to many different classes of compounds by complexation gas chromatography[6,12,15,19]. Yet, it may be postulated that that structural changes at the camphor moiety, e.g. by increasing the steric hindrance of pendant alkyl-substituents, will favorably alter the propensity toward chiral recognition. This reasoning commands interest also in the optimization of chiral paramagnetic "shift" reagents such as lanthanum(III)tris[3-(perfluoroacyl)-(1R)-camphorates] for the differentiation of enantiotopic nuclei in NMR spectroscopy[31,38]. Previous attempts at structural elaboration of the syn-C8-methyl group of camphor via 8-bromocamphor were not particularly successful[39]. We therefore applied

101

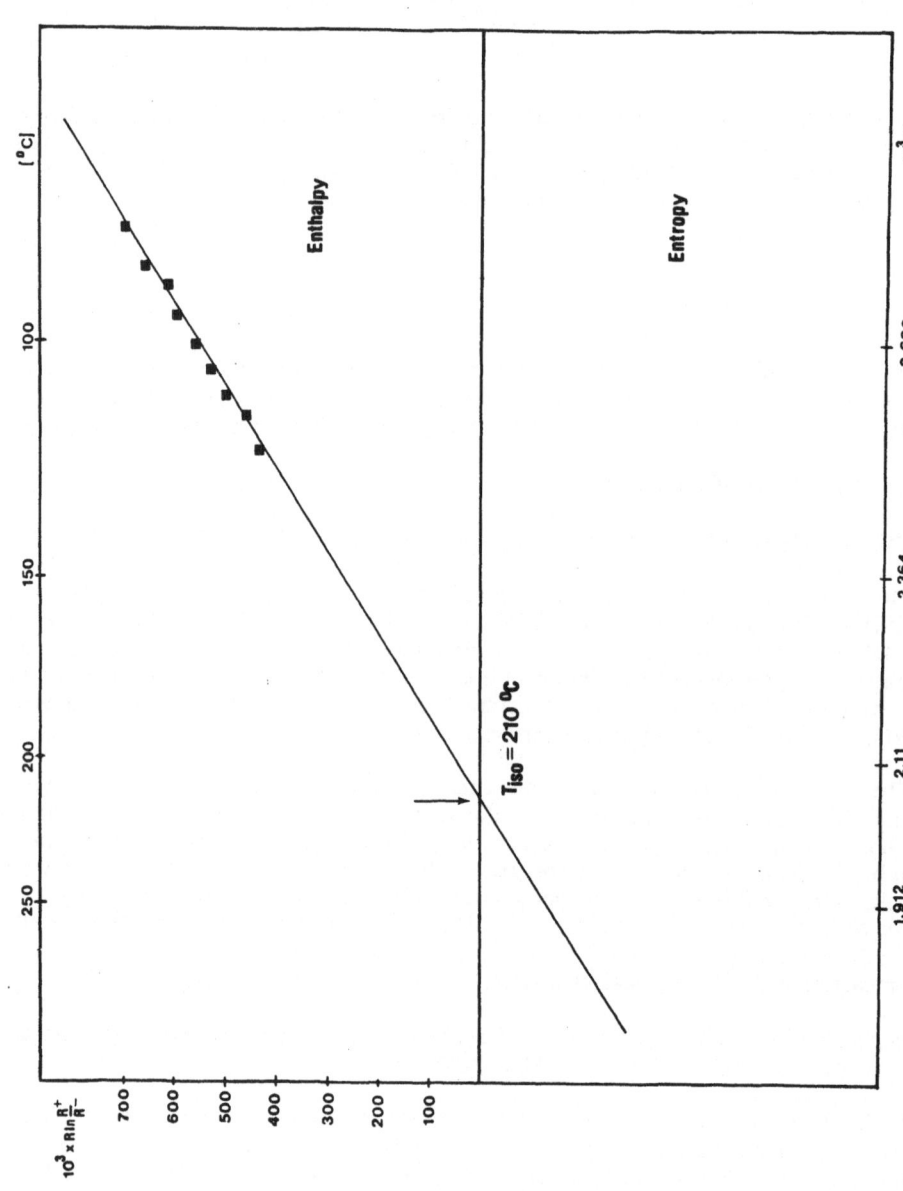

Scheme 2. The influence of enthalpy and entropy changes on enantiomer discrimination by complexation gas chromatography. The iso-enantioselective temperature T_{iso}. Tentative extrapolation for E-chalcogran (2).

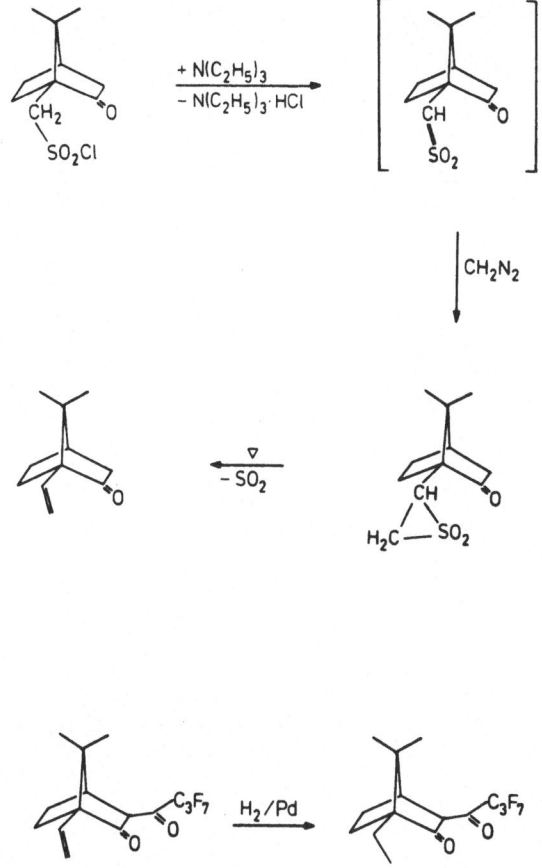

Scheme 3. Preparation of 10-methylene- and 10-methylcamphor.

the versatile transformation of camphor-10-sulfurylchloride via diazomethane into 10-methylenecamphor complemented by the subsequent hydrogenation to 10-methyl-camphor[32,37], cf. Scheme 3.

Acylation of 10-methylene- and 10-methyl-camphor with heptafluorobutanoylchloride to give 3-(heptafluorobutanoyl)-10-methylene-camphor and 3-(heptafluorobutanoyl)-10-methyl-camphor, respectively[31], are as straightforward as the preparation of the novel Chirametal specimens nickel(II)bis[3-(heptafluorobutanoyl)-(1S)-10-methylene-camphorate] (1b) and nickel(II)bis[3-(heptafluorobutanoyl)-(1R)-10-methyl-camphorate] (1c)[12]. It has also been found convenient to prepare the ligand 3-(heptafluorobutanoyl)-10-methyl-camphor by hydrogenation of 3-(heptafluorobutanoyl)-10-methylene-camphor[37].

The propensity of the three Chirametal stationary phases 1a, 1b, 1c towards enantiomer separation by complexation gas chromatography has been screened with 11 chiral solutes at 70°C and 90°C. The results are contained in Tables 2 and 3.

As a concise measure of enantioselectivity the quantity of $-\Delta_{R,S}(\Delta G°)$ rather then the chromatographic separation factor α, which is important only for practical reasons, is given therein. Inspection of the Tables 2 and 3 immediately reveals that 1c (cf. Fig. 2) displays a substantially higher enantioselectivity than the standard stationary phase 1a while the performance of 1b is either slightly improved or inferior to that of 1a. An expected increase of $-\Delta_{R,S}(\Delta G°)$ with increasing steric hindrance exerted by the alkyl-substituent is noted for oxiranes on 1c but not on 1a. On the latter Chirametal stationary phase, for

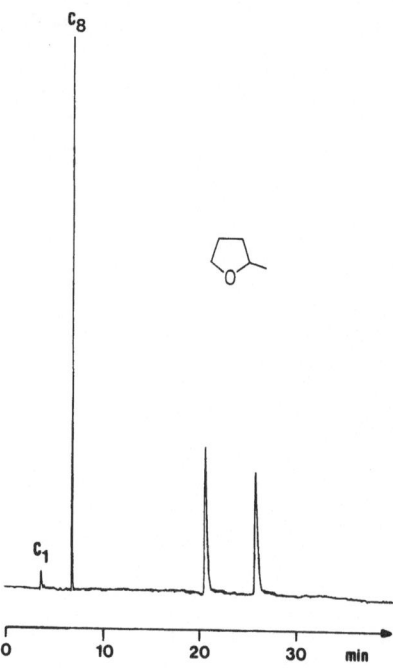

Fig. 2. Enantiomer separation of 2-methyltetrahydrofuran on a 37m x 0.25 mm I.D. glass capillary column coated with 0.12m nickel(II)*bis*[3-(heptafluorobutanoyl)-(1*R*)-10-methyl-camphorate] (1c) in OV-101. Oven temperature: 90°C; inlet-pressure: 1 bar nitrogen.

example, isopropyloxirane is not resolved[12]. Thus, 1c may be in the future replace 1a as a useful Chirametal stationary phase in complexation gas chromatography.

1b may preferentially be employed for enantiomer separation in selected instances, despite its weak performance for the chiral solutes listed in Tables 2 and 3. Thus, 1b has proved to be a highly enantioselective Chirametal stationary phase for carboxylic and 2-halocarboxylic esters (cf. Fig. 3)[40] and for isomeric roseoxides (cf. Fig. 4), respectively.

While the improved enantioselectivity of 1c as compared to 1a may be rationalized by the increase of steric constraints upon coordination of a solute, the odd behavior of 1b, containing an unsaturated entity flanking the camphor moiety, is not understood at present. The data in Tables 2 and 3 reveal another important aspect, namely, that $-\Delta_{R,S}(\Delta G°)$ is independent from R', that is, enantiomer discrimination is independent from the degree of coordination as has already been found for Z- and E-chalcogran 2 *vide supra*. The remarkable difference of the enantioselectivity exerted by 1a, 1b and 1c on chalcogran (Z- and E-2-ethyl-1,6-dioxaspiro[4,]nonane) 2 can be recognized from Fig. 5.

While 1c shows a highly improved enantioselectivity as compared to 1a, almost no enantiomer separation is observed with 1b, although chemical interaction is very pronounced as judged from the high retention-increase R'. Moreover, the order of elution of the enantiomers of 2 are reversed (peak-inversion) on 1b as compared to 1a and 1c! *

* *Note added in proof:* This phenomenon has been clarified as being due to a low isoenantioselective temperature T_{iso} (V. Schurig, F. Ossig, R. Link, *Augew. Chem. Int. Ed. Engl.*, in print (1989).

Previously, a quadrant rule has been formulated which correlates the order of elution of chiral alkylsubstituted oxiranes with their absolute configuration[41]. This rule is valid for all oxiranes and Chirametal stationary phases listed in Tables 1 and 2.

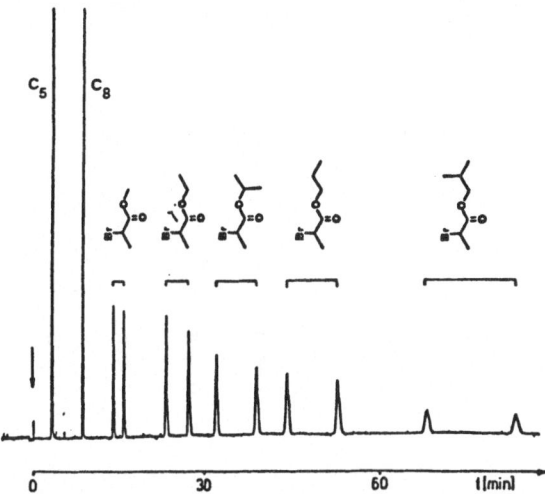

Fig. 3. Enantiomer separation of 2-bromo-carboxylic acid esters on a 28m x 0.25 mm I.D. glass column coated with 0.1m nickel(II)*bis*[3-(heptafluorobutanoyl)-(1S)-10-methylene-camphorate] (1b) in OV-101. Oven temperature: 90°C; inlet-pressure: 0.8 bar nitrogen[36].

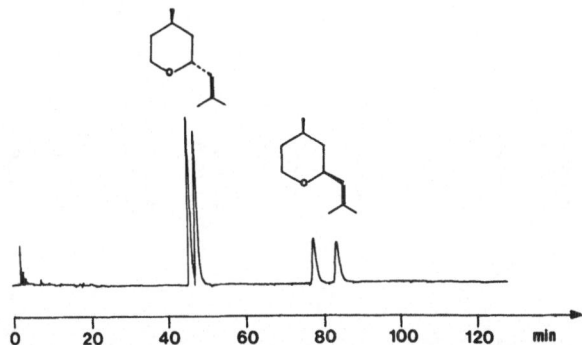

Fig. 4. Enantiomer separation of *cis*- and *trans*-tetrahydro-4-methyl-2-(2-methyl-1-propenyl)-pyrane (cis- and trans-roseoxide) on a 34m x 0.25 mm I.D. glass capillary column coated with 0.13m nickel(II)*bis*[3-(heptafluorobutanoyl)-(1S)-10-methylene-camphorate] (1b) in OV-101. Oven temperature: 90°C; inlet-pressure: 1 bar nitrogen.

Table 2. Retention-increase R' and Enantioselectivity -Δ$_{R,S}$(ΔG°) of 11 Solutes on the Chirametal Stationary Phases 1a, 1b, 1c in OV-101 at 70°C.

Solutes		R'	-Δ$_{R,S}$(ΔG°) [KJ/Mol]	R'	-Δ$_{R,S}$(ΔG°) [KJ/Mol]	R'	-Δ$_{R,S}$(ΔG°) [KJ/Mol]
Methyloxirane		77.26(3) 91.21(3)	0.47(3)	55.71(3) 66.93(8)	0.52(4)	97.78(8) 129.95(0)	0.81(1)
Ethyloxirane		83.58(8) 90.78(1)	0.23(5)	63.30(2) 67.85(0)	0.19(8)	90.23(5) 127.05(9)	0.97(6)
trans-2,3-Dimethyloxirane		36.27(8) 45.66(2)	0.65(6)	24.94(0) 28.03(8)	0.33(4)	27.52(6) 36.68(4)	0.81(9)
tert.-Butyloxirane		41.18(6) 44.95(1)	0.25(0)	30.47(5) 35.90(8)	0.46(8)	29.84(7) 46.63(1)	1.27(3)
Isopropyloxirane		68.24(9)	-	32.15(9)	-	72.33(0) 105.12(0)	1.06(6)
Epichlorohydrin		11.16(7) 13.33(9)	0.50(7)	8.86(4) 10.31(2)	0.43(2)	12.25(1) 18.31(4)	1.14(7)
Epibromohydrin		12.00(1) 14.20(4)	0.48(1)	9.83(0) 11.14(7)	0.35(9)	13.65(7) 20.01(1)	1.09(0)
2-Methyltetrahydrofuran		84.66(0) 94.72(4)	0.32(0)	51.54(8) 68.33(1)	0.80(4)	45.39(0) 64.75(2)	1.01(3)
tert.-Butylmethylcarbinol		42.34(4) 46.24(1)	0.25(1)	26.22(1) 26.97(0)	0.08(0)	28.52(7) 35.96(3)	0.66(1)
2-Bromopropionic acid ethyl ester		3.92(6) 4.22(8)	0.21(1)	2.84(0) 4.69(3)	1.43(3)	2.07(5) 2.64(7)	0.69(4)
Chalcogran	Z	1.59(1) 2.16(1)	0.87(3)	2.07(5) 2.21(9)	0.19(1)	2.06(1) 3.57(9)	1.57(4)
	E	4.73(3) 6.78(8)	1.03(0)	6.28(1)	-	7.30(6) 11.73(6)	1.35(2)

Table 3. Retention-increase R' and Enantioselectivity -ΔR,S(ΔG°) of 11 Solutes on the Chirametal Stationary Phases 1a, 1b, 1c in OV-101 at 90°C.

Solutes		R'	-ΔR,S(ΔG°) [KJ/Mol]	R'	-ΔR,S(ΔG°) [KJ/Mol]	R'	-ΔR,S(ΔG°) [KJ/Mol]
Methyloxirane		41.65(7) 47.15(7)	0.37(4)	33.63(7) 38.33(3)	0.39(4)	30.67(7) 38.33(3)	0.67(2)
Ethyloxirane		42.38(1) 44.55(6)	0.15(1)	34.40(8) 36.64(6)	0.19(0)	27.55(6) 34.77(4)	0.70(2)
trans-2,3-Dimethyloxirane		18.11(0) 21.17(8)	0.47(2)	14.36(2) 15.46(6)	0.22(4)	9.19(0) 11.22(7)	0.60(4)
tert.-Butyloxirane		20.74(1) 22.04(8)	0.18(4)	15.48(0) 17.47(2)	0.36(5)	8.87(0) 12.76(0)	1.09(8)
Isopropyloxirane		39.68(1)	-	20.19(2) 20.72(3)	0.07(8)	19.56(8) 26.26(8)	0.88(9)
Epichlorohydrin		5.97(8) 6.82(0)	0.39(8)	5.66(2) 6.32(4)	0.33(4)	4.03(9) 5.22(5)	0.77(7)
Epibromohydrin		6.56(1) 7.43(9)	0.37(9)	6.33(2) 6.97(0)	0.29(0)	4.72(9) 6.40(5)	0.91(6)
2-Methyltetrahydrofuran		29.75(0) 32.68(3)	0.28(4)	23.92(5) 30.05(9)	0.68(9)	11.79(4) 15.65(7)	0.85(5)
tert.-Butylmethylcarbinol		20.19(8) 21.65(4)	0.21(0)	13.35(4) 13.79(7)	0.09(9)	8.29(3) 9.98(8)	0.56(1)
2-Bromopropionic acid ethyl ester		2.01(0) 2.12(4)	0.16(7)	1.54(2) 2.33(4)	1.25(1)	0.96(8) 1.11(3)	0.42(1)
Chalcogran	Z	1.34(2) 1.71(2)	0.73(5)	1.81(4) 1.91(8)	0.16(8)	1.20(8) 1.84(6)	1.28(0)
	E	3.96(6) 5.31(7)	0.88(5)	5.30(8) 5.42(6)	0.06(6)	3.97(3) 5.87(9)	1.18(3)

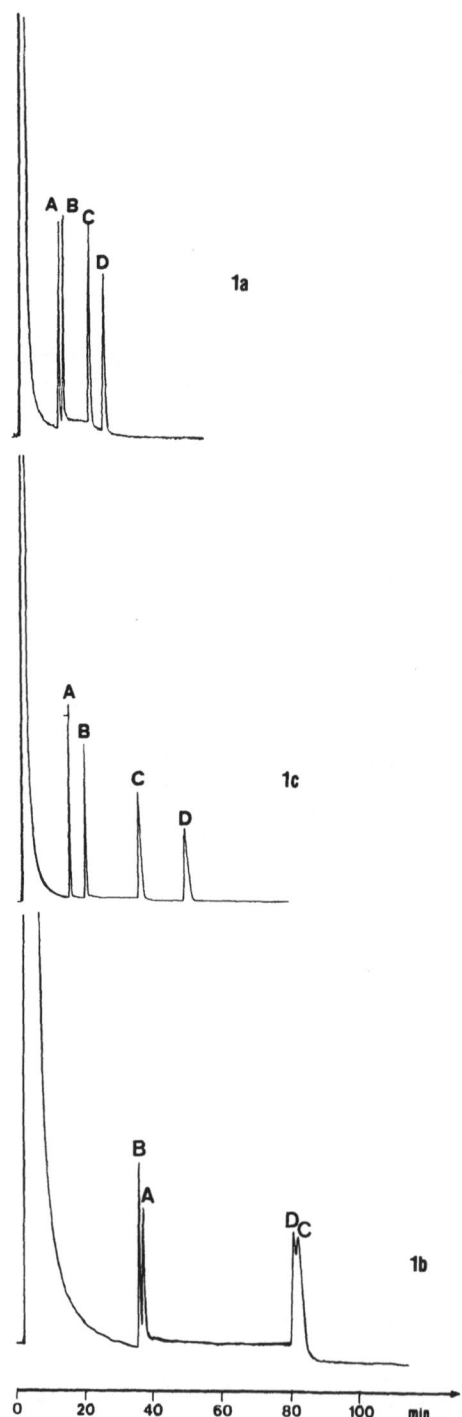

Fig. 5. Influence of structural changes at the C(10) position of camphor upon enantioselectivity of camphorato-Chirametal stationary phases 1a, 1b, 1c. Enantiomer
separation of chalcogran (Z- and E-2-ethyl-1.6-dioxaspiro[4.4]nonane) 2. Top: 37m
x 0.25 mm I.D. glass capillary column coated with 0.12m nickel(II)bis[3-
(heptafluorobutanoyl)-(1R)-10-methylene-camphorate] (1a) on SE-54 (0.2 μ). Oven
temperature: 80°C; inlet-pressure: 1 bar nitrogen. Center: 25m x 0.25 mm I.D. glass
capillary column coated with 0.13m nickel(II)bis[3-(heptafluorobutanoyl)-(1R)-10-
methylene-camphorate] (1c) on SE-54 (0.2 μ). Oven temperature: 80°C; inlet-

pressure: 1 bar nitrogen. Bottom: 24m x 0.25 mm I.D. glass capillary column coated with 0.125m nickel(II)*bis*[3-(heptafluorobutanoyl)-(1S)-10-methylene-camphorate] (1b) on SE-54 (0.2 μ). Oven temperature: 80°C; inlet-pressure: 1 bar nitrogen. Peak assignment: A = 2S,5S-2; B = 2R,5R-2; C = 2R,5S-2; D = 2S,5R-2[13].

3) *Synthesis of a Polymeric Chirametal Stationary Phase "Chirasil-Metal"[32]*

The coupling of the classical hydrogen-bonding selectors[1] such as the L-valine diamide phase[42] to polysiloxanes[2] was a milestone in gas chromatographic enantiomer separation. This approach combined the high enantioselectivity of the selector with the high thermal stability, high viscosity and involatility of polysiloxanes. Two strategies have been devised in the preparation of chiral polysiloxanes: (i) total synthesis by co-polymerization ("Chirasil-Val")[43], (ii) the "polymer-analogous reaction" of the selector and a polysiloxane with appropriate chemical functions[44,45].

Clearly, method (i) offers the possibility to tailor the desired properties of the polymer. In the attempt to link a Chirametal stationary phase to a polysiloxane matrix a number of strategies can be envisioned, e.g. the chiral selector may be attached via the acyl residue of the terpene ketone or via the terpene backbone to the polymer.

Recently, the anchoring of the chiral cyclopropanation catalyst copper(II)*bis*[3-(trifluoroacetyl)-(1R)-camphorate] on silica has been achieved via catalytic hydrosilylation of 10-methylene-camphor with trichlorosilane followed by the coupling of the silylated ligand to hydroxy groups of silica and reaction of the polymeric β-diketone with copper(II) or nickel(II)[46].

We have selected (3-heptafluorobutanoyl)-(1S)-10-methylene-camphor (note the formal change of the descriptor as compared to camphor due to the priority change in the sequence rule of Cahn, Ingold and Prelog) as a versatile ligand for the preparation of Chirasil-Metal stationary phases by the total synthesis strategy (i). The approach is outlined in Scheme 4.

Thus, in a first step (3-heptafluorobutanoyl)-(1S)-10-(dimethoxymethylsilyl)-methyl-camphor was prepared by H_2PtCl_6-catalyzed hydrosilylation of (3-heptafluorobutanoyl)-(1S)-10-methylene-camphor with methyldimethoxysilane (the use of chlorosilanes is avoided because hydrochloric acid formed in subsequent hydrolysis may decompose the metal-β-diketone) which proceeds quantitatively in the desired anti-Markovnikov mode of addition. The reaction was monitored by proton nuclear magnetic resonance spectroscopy (disappearance of $=CH_2$, appearance of -SiMe and -SiOMe proton resonances). Prior to polymerization, (3-heptafluorobutanoyl)-(1S)-10-(dimethoxymethylsilyl)-methylcamphor was converted into its anion with sodium hydride and reacted with $NiCl_2$[12] to give the silylated Chirametal monomer. The hydrolysis of the monomer was carried out in methanol/water. Adduct formation of the nickel-β-diketone with H_2O can be tolerated at this step. The co-monomer diethoxymethylvinylsilane was hydrolyzed at elevated temperatures via its organosilandiol. Co-polymerization of hydrolyzed nickel(II)bis-[(3-heptafluorobutanoyl)-(1S)-10-(dimethoxymethylsilyl)-methylcamphorate] and hydrolyzed diethoxymethylvinylsilane (1:4.5) in the presence of trimethylsilanol (in order to regulate the molecular weight of the polymer) was carried out with tetramethylammoniumhydroxide as catalyst. Residual silanol groups in the polymer were removed by end-capping with 1.3-divinyltetramethyldisilazane. The Chirasil-Metal 3 polymer was characterized by infrared-spectroscopy showing characteristic bands for the Si-O-Si-group between 1000 and 1100 cm^{-1}. The nickel content has been determined by atomic absorption spectroscopy to 1.1% corresponding to a ratio of 1:7 of the monomer units. The ultra-violet-spectrum of the Chirasil-Metal 3 polymer showed a band at λ_{max} = 322 nm characteristic for the π–π*-transition of the metal-chelate carbonyl groups.

In the preparation of the Chirasil-Metal 3 polymer the nickel(II) β-diketone was exposed to strongly coordinating solvents and catalysts such as water and TMAH. It is therefore anticipated that coordination-sites required for the selectand-selector association are blocked via adduct formation. Indeed, the polymer did not exert retention-increase R' upon gas

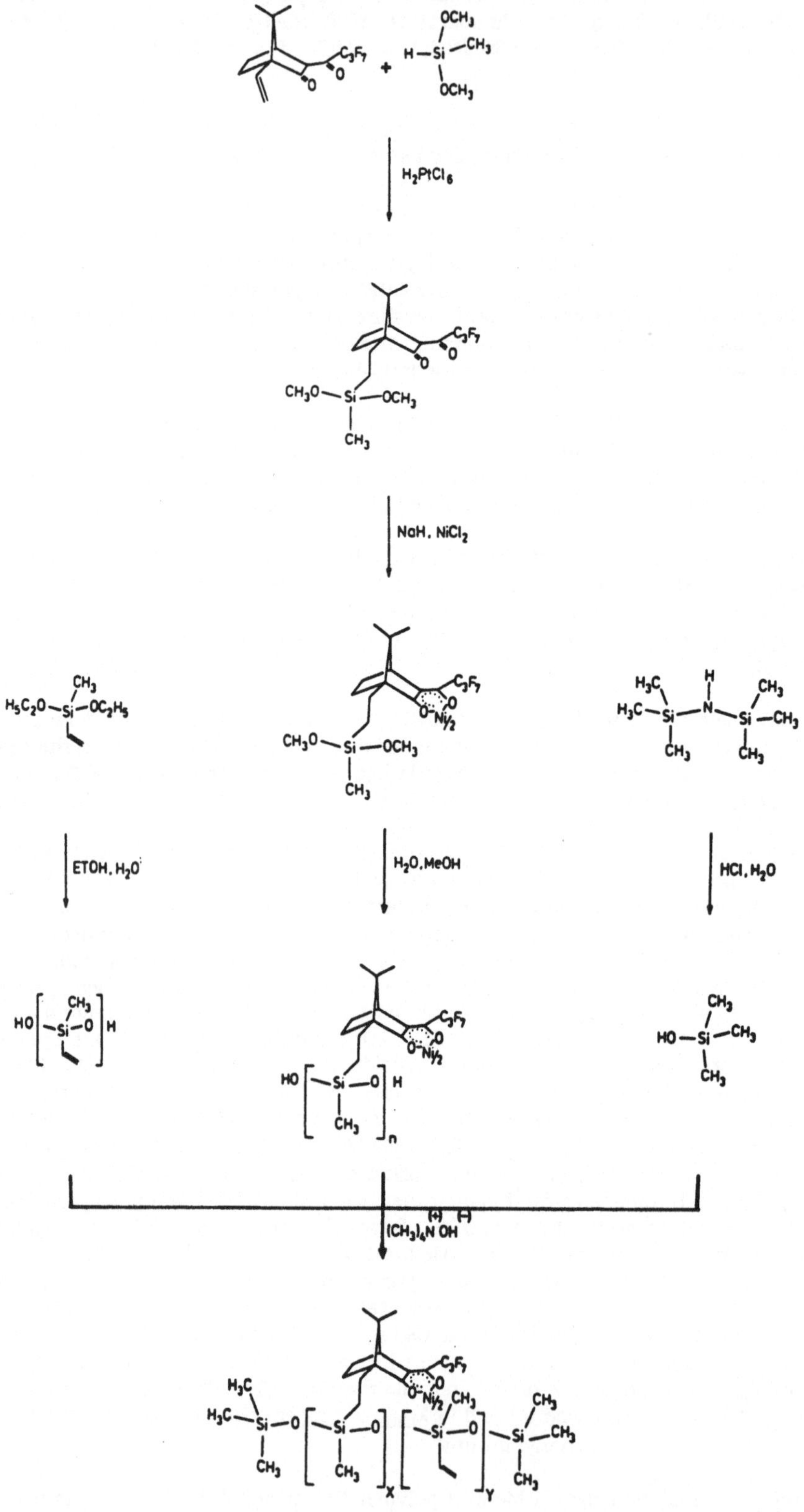

Scheme 4. Preparation of the nickel(II) Chirasil-Metal 3-stationary phase.

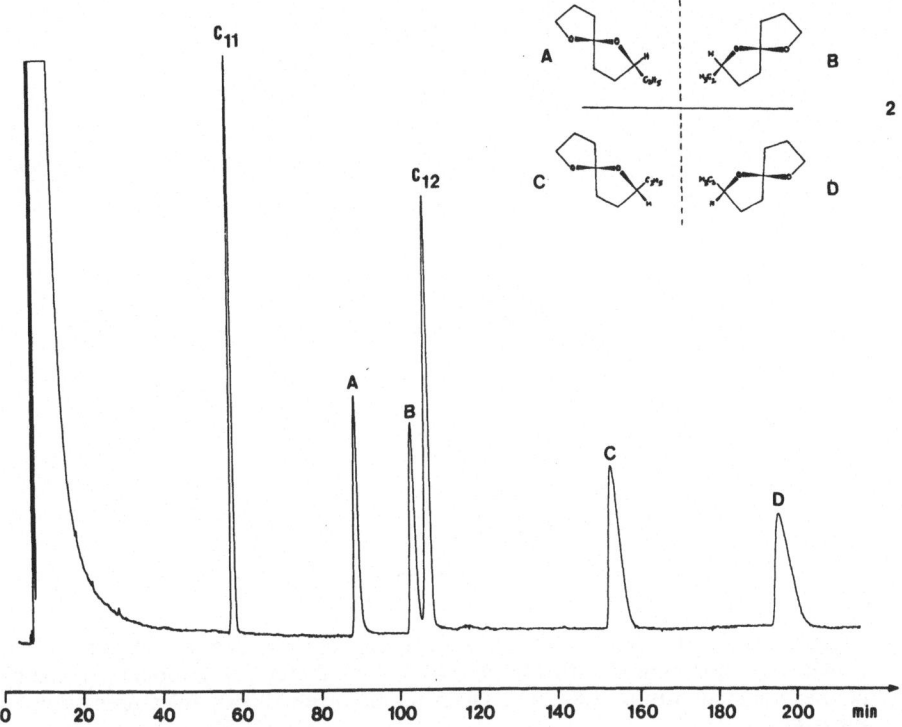

Fig. 6. Enantiomer separation of chalcogran (Z- and E-2-ethyl-1.6-dioxaspiro[4.4]nonane) 2 on a 37m x 0.25 mm I.D. glass capillary column coated with the nickel(II)-Chirasil-Metal 3 stationary phase (ratio of the monomer units: 4.5 : 1). Oven temperature: 90°C; inlet-pressure: 1 bar nitrogen.

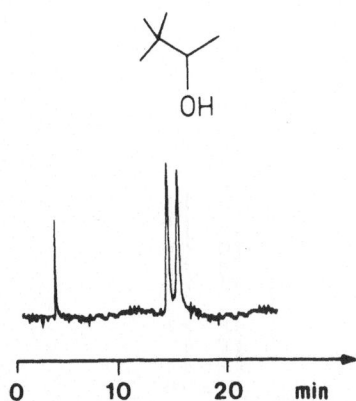

Fig. 7. Enantiomer separation of 3.3-dimethylbutan-2-ol (tert-butylmethyl-carbinol) on a 37m x 0.25 mm I.D. glass capillary column coated with the nickel(II)-Chirasil-Metal 3 stationary phase (ratio of the monomer units: 4.5 : 1). Oven temperature: 90°C; inlet-pressure: 1 bar nitrogen.

chromatography of, e.g. oxiranes after a 12 h conditioning period at 70-80°C. However, after conditioning for 4 h at 190°C the Chirasil-Metal stationary phase was capable of resolving racemic compounds by complexation gas chromatography. Typical chromatograms are shown in Figs. 6-9.

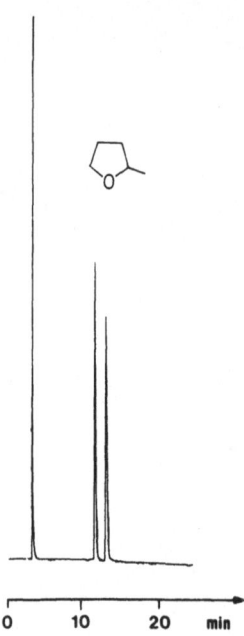

Fig. 8. Enantiomer separation of 2-methyltetrahydrofuran on a 37m x 0.25 mm I.D. glass capillary column coated with the nickel(II)-Chirasil-Metal 3 stationary phase (ratio of the monomer units: 4.5 : 1). Oven temperature: 90°C; inlet-pressure: 1 bar nitrogen.

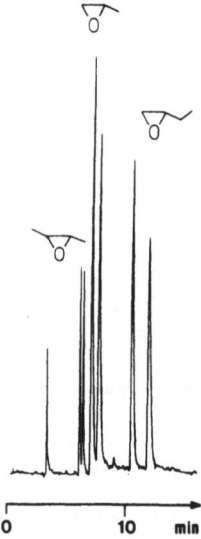

Fig. 9. Enantiomer separation of *trans*-2.3-dimethyloxirane, methyl- and ethyloxirane on a 37m x 0.25 mm I.D. glass capillary column coated with tne nickel(II)-Chirasil-Metal 3 stationary phase (ratio of the monomer units: 4.5 : 1). Oven temperature: 90°C; inlet-pressure: 1 bar nitrogen.

The absence both of "peak-coalescence of the second kind" for chalcogran 2[47] (Fig. 6) and of peak-tailing for tert-butylmethylcarbinol (Fig. 7) is indicative of the highly deactivated state of the polymer coated to the glass open-tubular column.

A test on the temperature stability of the Chirametal stationary phases 1a, 1b, 1c and the Chirasil-Metal 3 stationary phase at 140°C for 168 h revealed the following decrease of the retention-increase R' for ethyloxirane (checked at 90°C):

1a: 17.3%
1b: 7.6%
1c: 26.0%
Chirasil-Metal: approximately 4.8%.

Thus, the polymeric stationary phase showed the highest thermal stability, although unfortunately, some loss of activity in respect to chemical interaction has also been noted. Thus, at the present state of the art, high thermal stability of Chirasil-Val stationary phases has not yet been achieved.

The Chirasil-Metal 3 stationary phase contains pendant vinyl groups which were introduced intentionally. Immobilization of the Chirasil-Metal 3 polymer by cross-linking as well as the permanent attachment to glass (or fused silica) surfaces commands interest in respect of non-extractical stationary phases in gas chromatography and super-critical-fluid chromatography. These topics are currently being pursued in our laboratory.

Acknowledgements

Generous support of this work by "Deutsche Forschungsgemeinschaft" and "Fonds der chemischen Industrie" is thankfully acknowledged.

REFERENCES

1.   E. Gil-Av, B. Feibush and R. Charles-Sigler, *Tetrahedt.Lett.*, 1009 (1966); E. Gil-Av, J.Mol.Evol., 6:131 (1975).
2.   H. Frank, G. J. Nicholson and E. Bayer, *Angew.Chem.Int.Ed.Engl.*, 17:363 (1978).
3.   W. A. König, "The Practice of Enantiomer Separation by Capillary Gas Chromatography," Hüthig, Heidelberg (1987).
4.   V. Schurig, *Angew.Chem.Int.Ed.Engl.*, 23:747 (1984).
5.   V. Schurig, "Asymmetric Synthesis," (J.D. Morrison, ed.), Academic Press, New York, vol. 1, p. 59 (1983).
6.   V. Schurig, *Kontakte (Darmstadt)* 3 (1986).
7.   V. Schurig, *Chromatographia*, 13:263 (1980).
8.   V. Schurig, *Angew.Chem.Int.Ed.Engl.*, 16:110 (1977).
9   V.Schurig and E. Gil-Av, *Isr.J.Chem.*, 15:96 (1976/77).
10. V. Schurig and W. Bürkle, *Angew.Chem.Int.Ed.Engl.*, 17:132 (1978).
11. V. Schurig and R. Weber, *J.Chromatogr.*, 217:51 (1981).
12. V. Schurig and W. Bürkle, *J.Am.Chem.Soc.*, 104:7573 (1982).
13. V. Schurig and R. Weber, *J.Chromatogr.*, 289:321 (1984).
14. V. Schurig, U. Leyrer and R. Weber, *J.High Res.Chromatogr.*, *Chromatogr.Comm.*, 8:459 (1985).
15. R. Weber and V. Schurig, *Naturwissenschaften*, 71:408 (1984).
16. D. Wistuba and V. Schurig, *Angew.Chem.Int.Ed.Engl.*, 25:1033 (1986).
17. A. Mosandl, G. Heusinger, D. Wistuba and V. Schurig, *Z.Lebensm.Unters.Forsch.*, 179:385 (1984).
18. V. Schurig, U. Leyrer and U. Kohnle, *Naturwissenschaften*, 72:211 (1985).
19. V. Schurig, "Bioflavour '87," (P. Schreier, ed.), Walter de Gruyter, Berlin (1988).
20. G. Singer, G. Heusinger, O. Fröhlich, P. Shreier and A. Mosandl, *J.Agric.Food Chem.*, 34:1029 (1986).
21. W. Kirmse, P. V. Chiem and V. Schurig, *Tetrahedr.Lett.*, 26:197 (1985).
22. K. Keinan, K. K. Seth and R. Lamed, *J.Am.Chem.Soc.*, 108:3474 (1986).
23. P. Werkhoff and R. Hopp, "Progress in Essential Oil Research," (E.-J. Brunke, ed.), Walter de Gruyter & Co., Berlin-New York, 529 (1986).
24. R. M. Carman, I. C. MacRae and M. V. Perkins, *Aust.J.Chem.*, 39:1739 (1986).
25. R. L. Halterman, W. R. Roush and L. K. Hoong, *J.Org.Chem.*, 52:1152 (1987).
26. B. Koppenhoefer, Thesis, University of Tübinger, FRG (1980).

27. B. Koppenhoefer and E. Bayer, *Chromatographia*, 19:123 (1984).
28. R. Weber, Thesis, University of Tübingen (1983).
29. K. Grob, "Making and Manipulating Capillary Columns for Gas Chromatography," Hüthig, Heidelberg, p.124, 1707 (1986).
30. R. Weber, K. Hintzer and V. Schurig, *Naturwissenschaften*, 67:453 (1980).
31. M. D. McCreary, D. W. Lewis, D. L. Wernick and G. M. Whitesides, *J.Am.Chem.Soc.*, 96:1038 (1974).
32. N. Fischer and G. Opitz, *Org.Synth., Coll.* Vol. 5 877 (1973).
33. R. O. Sauer, *J.Am.Chem.Soc.*, 66 (1944).
34. V.Schurig, E. Gil-Av, Isr.*J.Chem., Suppl. Proc.Isr.Chem.Soc.*, 9:220 (1971).
35. V. Schurig, R. C. Chang, A. Zlatkis and B. Feibush, *J.Chromatogr.*, 99:147 (1974).
36. W. Francke, G. Hindorf and W. Reith, *Angew.Chem.Int.Ed.Engl.*, 17:862 (1978).
37. R. Link, Thesis, University of Tübingen, (1987).
38. V. Schurig, *Tetrahedr.Lett.*, 3297 (1972).
39. R. Weber, Diploma Thesis, University of Tübingen (1979).
40. V. Schurig, A. Ossig and R. Link, *J.High Res.Chromatogr., Chromatogr.Comm.*, 11:89 (1988).
41. V. Schurig, B. Koppenhöfer and W. Bürkle, *Angew.Chem.Int.Ed.Engl.*, 17:937 (1978).
42. B. Feibush, *J.Chem.Soc., Chem.Comm.*, 544 (1971).
43. H. Frank, G. J. Nicholson and E. Bayer, *J.Chromatogr.Sci.*, 15:174 (1977).
44. T. Saeed, P. Sandra and M. Verzele, *J.Chromatogr.*, 186:611 (1979).
45. W. A. König and I. Benecke, *J.Chromatogr.*, 209:91 (1981).
46. S. A. Matlin, W. J. Lough, L. C. Chan, D. M. H. Abram and Z. Zhou, *J.Chem.Soc., Chem.Comm.*, 1039 (1984).
47. V. Schurig, *J.Chromatogr.*, 441:135 (1988).

A *NOTE ON* USE OF VARIOUS COMMERCIALLY AVAILABLE CHIRAL STATIONARY PHASES IN SUPERCRITICAL FLUID CHROMATOGRAPHY

P. Macaudiere, M. Caude, R. Rosset and T. Tambute*

Laboratoire de Chimie Analytique de l'Ecole Supérieure de Physique et Chimie de Paris, 10 rue Vauquelin, 75231 Paris Cedex 05, France and *Direction des Recherches et Etudes Techniques, Centre d' Etudes du Bouchet, BP no 3, Le Bouchet , 91710 Vert-le-Petit, France

INTRODUCTION

In the past few years, there has been much interest in supercritical fluid chromatography (SFC)[1-3], due to the fact that SFC offers a number of advantages over liquid chromatography (LC) such as shorter analysis time, cheaper solvents and the use of detectors akin to those for gas chromatography. At the same time, there has been much interest in the field of chiral separations[4], exemplified by the development of many chiral stationary phases (CSPs), many of which are now commercially available.

In previous papers[5,6] we have reported the chiral resolution of phosphine oxides and amides on a Pirkle-type CSP, the (R)-N-(3,5-dinitrobenzoyl)phenylglycine. We have obtained similar retentions, selectivities and efficiencies when using sub- or supercritical fluid chromatography (SubFC or SFC) and LC. However, resolution per unit time was always greater with SubFC (or SFC) conditions rather than LC ones[6].

In this present work, three commercially available CSPs (Cyclobond I), Chiralcel OB and Chiralpak OT(+)) have been evaluated under SubFC and SFC conditions. For all these CSPs, a brief comparison of retention, selectivity, resolution per unit time and chiral recognition mechanisms between LC and SFC has been made.

EXPERIMENTAL

Apparatus

The apparatus for SFC has been described previously[1]. Carbon dioxide, kept in a container with an eductor tube, was passed into a Model 303 pump (Gilson, Villiers-le-Bel, France) through an ethanol cooling-bath. The pump head (10SC) was cooled in order to improve pump efficiency. The inlet adaptor and the cooling jacket were laboratory-made. Polar modifiers were added using a second Gilson pump and mixed with carbon dioxide in a Gilson mixer (Model 811). A constant-temperature water-bath provided temperature control for the column.

A polychrom 9060 diode-array detector (Varian, Palo Alto, C.A. U.S.A.) set at 229 nm was used without modification. The pressure was monitored by a manual back-pressure regulator (TESCOM, Model 26-1700; G.E.C. Composants, Asnières, France) connected in-line after the detector and maintained at 35° by a water-bath. All results were recorded with a

115

Shimadzu CR 3A integrator (Touzart et Matignon, Vitry-sur-seine, France). The standard operating conditions were: average pressure, 150 bar; temperature 25°C; average carbon dioxide flow-rate; 5 ml/min at - 5°C

For LC, we used a modular liquid chromatograph (Gilson). The standard operating conditions were; flow-rate, 1 ml/min; temperature, 25°C

Chiral Stationary Phases

Three columns were used: (1) a β–cyclodextrin-bonded column (cyclobond I, Advanced Separation Technologies, Prolabo, Paris, France), particle size 5 μm (25 cm x 4.6 mm I.D.); (2) a cellulose tribenzoate adsorbed on macroporus silica (Chiralcel OB, Daicel Chemical Industries, Sochibo, Velizy-Villacoublay, France), particle size 10 μm (25 cm x 4.6 mm I.D.); (3) a (+)-polytriphenylmethyl methacrylate adsorbed on macroporous silica gel (Chiralpak OT(+), Daicel Chemical Industries), particle size 10 μm (25 cm x 4.6 mm I.D.).

Mobile Phase, Reagents and Solutes

Carbon dioxide was N 45-grade (99.995% pure) (Air Liquide, Alphagaz, Paris, France). Methanol, ethanol and 2-propanol of analytical grade and n-hexane of Lichrosolv grade were obtained from E. Merck (Darmstadt, F.R.G.).

The synthesis of amide[7] phosphine oxide[8] and α–methylene γ–lactone[9] samples have been described elsewhere. The albendazole sulfoxide was a gift from Professor Delatour (Ecole Nationale Veterinaire, Lyon, France).

RESULTS AND DISCUSSION

β–cyclodextrin Bonded CSP

A subcritical carbon dioxide mobile phase used in conjuction with methanol was effective for the resolution of racemic amides and phosphine oxides[10] on Cyclobond I CSP. This is the first example of a chiral separation obtained in the normal phase mode for this type of CSP. It has been possible to extend this method to normal phase liquid chromatography (NPLC). A comparison between LC and SubFC is given in Fig. 1. These results are somewhat surprising since the chiral recognition mechanism is based on inclusion complex formation between the hydrophobic moiety of the solute and the internal cavity of the β-cyclodextrin molecule (β-CD)[11]: a more common mobile phase would be a mixture of water and methanol (reversed-phase or RPLC) since apolar solvents such as hexane occupy the β–CD cavity and cannot easily be displaced by solutes.

Our chromatographic results can be explained, for SubFC, by the small size of the hydrophobic component in the mobile phase (i.e. carbon dioxide). This suggests that it could be displaced more easily than other apolar solvents from the cavity and/or that inclusion could happen even with a carbon dioxide molecule inside the cavity. The fact that selectivities are always greater in SubFC than in NPLC (for instance the β–naphthyl phosphine oxide remains unresolved in LC) is in agreement with the hypothesis that the inclusion process is favoured when carbon dioxide is used instead of hexane.

Since it has been impossible to separate the solute presented on Fig. 1 under RPLC conditions, it appears that SubFC (or NPLC) and RPLC are two complementary analytical techniques.

Cellulose Tribenzoate CSP

Cellulose tribenzoate coated on macroporous silica gel showed excellent capabilities of resolution of various enantiomers[12]. It was interesting to test this CSP (Chiralcel OB) under sub- or supercritical fluid chromatographic conditions since the chiral selector is not grafted to silica gel but coated on it. We have demonstrated the possibility of work at a pressure of up to 200 bar without loss of chromatographic performance of the silica bed: a

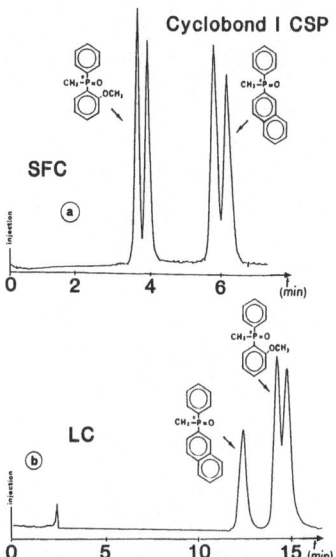

Fig. 1. Comparison of the resolution of the 2-naphthyl and o-anisyl phosphine oxide enantiomers on Cyclobond I. (a) SubFC: flow-rate, 4.5 ml/min; mobile phase, carbon dioxide-methanol (92:8, w/w); temperature, 25°C; average column pressure, 150 bar; detection, 234 nm. (b) LC: flow-rate, 1 ml/min; mobile phase, hexane-ethanol (85:15, v/v):temperature, 25°C; detection, 234 nm.

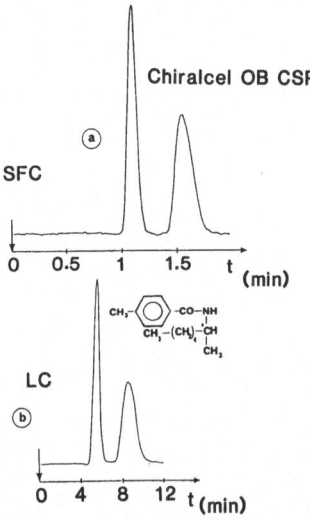

Fig. 2. Comparison of the resolution of the amide derivative of aminoheptane enantiomers on Chiralcel OB. (a) SFC: flow-rate, 5 ml/min; mobile phase, carbon dioxide-2-propanol (90:10, w/w); temperature, 35°C average column pressure, 130 bar; detection, 229 nm. (b) LC: flow-rate, 1 ml/min; mobile phase, hexane-2-propanol (90:10, v/v); temperature, 35°C detection, 230 nm.

slight decrease of efficiency (15%) has been observed during the first hours of use but selectivities remained constant during all the study.

Spectacular results have been observed with subcritical mobile phases: for enantiomeric amides used as test solutes by Wainer et al[13], the analysis times were reduced and in Fig. 2, for instance, it was reduced to 1/7th whilst keeping a constant resolution of 2 and a

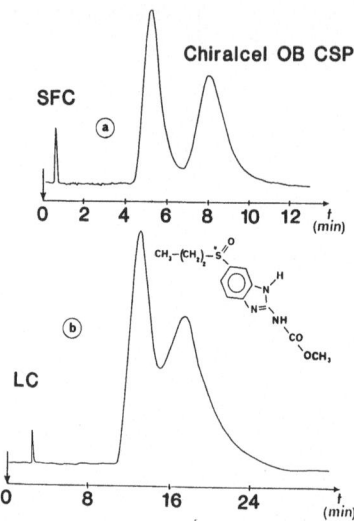

Fig. 3. Comparison of the resolution of the albendazole sulfoxide enantiomers on Chiralcel OB. (a) SubFC: flow-rate, 5 ml/min; mobile phase, carbon dioxide-2-propanol (93:7, w/w); temperature, 25°C; average column pressure, 140 bar; detection, 220 nm. (b) LC: flow-rate, 1 ml/min; mobile phase, hexane-ethanol (90:10, v/v); temperature, 25°C; detection, 220 nm.

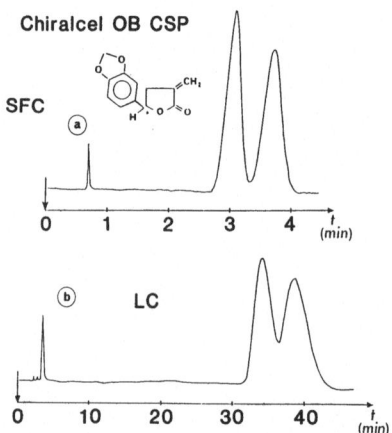

Fig. 4. Comparison of the resolution of the α–methylene γ–lactone enantiomers on Chiralcel OB. (a) SubFC: flow-rate, 5 ml/min; mobile phase, carbon dioxide-2-propanol (92:8, w/w); temperature, 25°C; average column pressure, 140 bar; detection, 229 nm. (b) LC: flow-rate, 1 ml/min; mobile phase, hexane-ethanol (85:15, v/v); temperature, 25°C; detection, 230 nm.

constant selectivity of 1.9. For other solutes, cf albendazole sulfoxide (Fig. 3) or α–methylene γ–lactone (Fig. 4) separations obtained by SubFC were impossible to achieve with LC, even with an analysis time ten times greater (Fig. 4). It is important to note that only subcritical fluid chromatography provides a valuable analytical technique for the control of the enantiomeric purity of these two drugs.

Unlike Pirkle-type CSPs[6], some differences in selectivities are observed and are under investigations. These results, in conjunction with the study of the influence of the nature of polar modifiers on retention and selectivity, will permit a better understanding of the chiral recognition mechanism, which seems to be a complex combination of hydrogen bonding, π–π and amide dipole interactions, and inclusion complex formation[12-14].

(+)-Polytriphenylmethyl Methacrylate CSP

The Chiralpak OT(+) consists of a macroporous silica gel coated with low-molecular weight (+)-polytriphenylmethylmethacrylate ((+)-PTrMA) whose chirality is due only to helicity. This CSP has great potentialy to separate various enantiomers, especially those containing aromatic groups[15]. In LC, methanol and acetonitrile are commonly used as mobile phases. In SFC, however, mixtures of carbon dioxide and various polar modifiers (content generally lower than 10%) are used. At higher contents, the diffusion coefficient, D_m, tends toward that of a liquid phase[16] and the benefit of the CSP is lost.

The change in mobile phase nature (pure methanol or carbon dioxide-methanol 92:8) explains the great differences in retention and selectivities between LC and SubFC (Fig. 5). These differences also depend on the nature of the solute too. For instance, the α–methylene γ–lactone selectivity remains similar (1.34 in SubFC; 1.44 in LC) while that of bi-naphthol is drastically changed (1.42 in SubFC; 2.52 in LC). At the same time, the elution orders are reversed, the bi-naphthol being eluted first in LC and second in SubFC.

All these results seem to indicate that two different sites are involved in the chiral recognition process: chiral sites located on a single polymer chain and chiral cavities formed by adjacent chains. Okamoto et al[15] have already described this phenomenon and they have favoured one type of site (the chiral cavities) by increasing the (+)-PTrMA/silica gel ratio (and thus the number of side chain interactions): they have reported that the selectivity values could be increased (for bi-naphthol for instance) or could present a maximum before decreasing. In our case, chiral cavity influence due to adjacent polymer chains seems to disappear in SubFC leading to low values of selectivity for bi-naphthol. Two explanations can be advanced: polymer chains are better solvated in carbon dioxide (swelling phenomenon) or, more probably, carbon dioxide molecules hinder the cavities and prevent the solute from interacting with them.

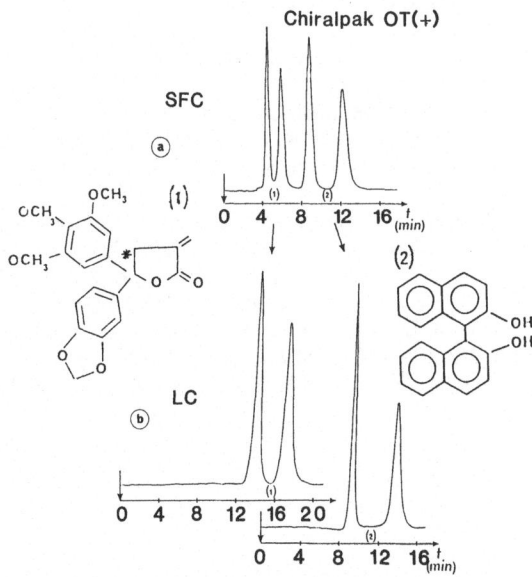

Fig. 5. Comparison of the resolution of bi-naphthol and α–methylene γ-lactone enantiomers on Chiralpak OT(+). (a) SubFC: flow-rate, 5 ml/min; mobile phase, carbon dioxide-methanol (92:8, w/w); temperature, 22°C; average column pressure, 140 bar; detection, 229 nm. (b) LC: flow-rate, 0.5 ml/min; mobile phase; methanol; temperature, 22°C; detection, 230 nm.

Other results confirm these hypotheses: the bi-naphthol selectivity increases from 1.39 to 1.44 when the methanol content in carbon dioxide varies from 4 to 16% whilst at the same time the α–methylene γ–lactone selectivity slightly decreases from 1.32 to 1.31.

As a conclusion, the Chiralpak OT(+) CSP can be used in SubFC but, due to the decrease of polarity in mobile phase, expected gains on analysis time are not observed: the higher velocity in SubFC (corresponding however to the same *reduced* velocity because of higher D_m value in SubFC than in LC) compensates only for the loss of polarity.

CONCLUSIONS

All the main NPLC chiral stationary phases can be used under sub- or supercritical fluid chromatographic conditions. For Pirkle-type[6] and for Chiralcel OB CSPs, lower retentions and higher resolutions were observed. This has not been the case for the Chiralpak OT(+) since this CSP is normally used with pure methanol as mobile phase.

Moreover, the feasibility of chiral separations in the normal phase mode with the Cyclobond I CSP has been demonstrated as well as the superiority of SubFC over NPLC. This made possible, for instance, the resolution of o-anisyl phosphine oxide enantiomers.

We have also shown that a systematic comparison of the results obtained in LC and in SFC allows a better understanding of chiral recognition mechanisms, especially when inclusion complexes are formed during the recognition process (Cyclobond I, Chiralcel OB and Chiralpak OT(+)).

REFERENCES

1. P. Mourier, P. Sassiat, M. Caude and R. Rosset, *Analusis*, 12:229 (1984).
2. P. Mourier, P. Sassiat, M. Caude and R. Rosset, *J.Chromatogr.*, 359:61 (1986).
3. *J.High Resolut.Chromatogr.Chromatogr.Commun.*, 3:136-85 (special issue) (1986).
4. Special issue on "Optical Resolution by Liquid Chromatography," S. Hara and J. Cazes, eds., *J.Liq.Chromatogr.*, 9:241-695 (1986).
5. P. Mourier, E. Eliot, A. Tambute, M. Caude and R. Rosset, *Anal.Chem.*, 57:2819 (1985).
6. P. Macaudiere, A. Tambute, M. Caude, R. Rosset, I. W. Wainer and M. C. Alembik, *J.Chromatogr.*, 371:177 (1986).
7. I. W. Wainer and M. C. Alembik, *J.Chromatogr.*, 367:59 (1986).
8. P. Pescher, M. Caude, R. Rosset and A. Tambute, *J.Chromatogr.*, 371:227 (1986).
9. C. Belaud, C. Roussakis, Y. Letourneux, N. El Alami and J. Villieras, *Synth.Comm.*, 15:1233 (1985).
10. P. Macaudiere, M. Caude, R. Rosset and A. Tambute, *J.Chromatogr.*, 405:135 (1987).
11. D. W. Armstrong, W. Demond and B. P. Czech, *Anal. Chem.*, 57:481 (1985).
12. T. Shibata, I. Okamoto and K. Ishii, *J.Liq.Chromatogr.*, 9:313 (1986).
13. I. W. Wainer and M. C. Alembik, *J.Chromatogr.*, 358:85 (1986).
14. E. Francotte, R. M. Wolf, D. Lohmann and R. Mueller, *J.Chromatogr.*, 347:25 (1985).
15. Y. Okamoto and K. Hatada, *J.Liq.Chromatogr.*, 9:369 (1986).
16. P. Sassiat, P. Mourier, M. Caude and R. Rosset, *Anal.Chem.*, 59:1164 (1987).

STRATEGIES FOR OPTIMISING CHIRAL SEPARATIONS

IN DRUG ANALYSIS

Anthony F. Fell and Terence A. G. Noctor

Department of Pharmaceutical Chemistry
University of Bradford, Bradford BD7 1DP, UK

SUMMARY

The principal strategies available for optimising chiral separations based on stereoselective column HPLC are briefly reviewed. The application of two particular approaches, response surface mapping and the modified sequential simplex method, is discussed in the context of optimising the separation of the enantiomers of oxamniquine on a first-generation α_1-acid glycoprotein phase. The performance of these approaches is discussed, and the potential contribution of novel chiral detector technology is considered with reference to more fully automated optimisation schemes in chiral HPLC.

INTRODUCTION

Chiral Aspects of Drug Activity

The chirality of a xenobiotic compound can have an impact on any of the processes of drug handling from absorption through to elimination. In general, the less specific the process, for example simple passive absorption, the more remote is the likelihood of any stereoselectivity being exhibited. Conversely, processes such as binding to postulated receptor sites and enzyme-mediated biotransformation, that rely heavily on the stereo-spatial nature of the substrate, quite commonly yield evidence of enantiomeric differences. This will result in compounds showing different pharmacokinetic or pharmacodynamic profiles. When enantiomers exhibit different degrees of pharmacological activity, the active enantiomer is termed the eutomer, whilst the lesser, or inactive isomer is termed the distomer[1]. The ratio of the potencies of the enantiomers is referred to as the 'eudismic ratio', and is a measure of the importance of stereoselectivity in the action of a particular drug. The logarithm of the eudismic ratio is known as the eudismic index, and is related to the difference in the binding energies of the enantiomers[1]. In certain cases enantiomers may have entirely different biological activities, which may in fact lead to 'side-effects' being observed for the racemate. For example, (+)-ketamine is a hypnotic and analgesic drug and adverse effects are known to be associated with the (-)-enantiomer[2].

Stereoselective Separations

Conventional methods for enantiomeric analysis are not easily applicable to routine studies. By far the most widely used method has been separation through the formation of diastereoisomers. This requires the availability of a suitable chiral reagent, which is completely optically pure. Even if such a reagent is obtained, further problems may arise, in the form of kinetic fractionation, whereby, due to differences in energies of intermediates for the two enantiomers, the final ratio of products may not necessarily reflect accurately the original ratio of enantiomers. This is obviously a major disadvantage in quantitative work. The ability to analyze enantiomers as such, or direct resolution (the previously described method being indirect resolution), is therefore to be preferred.

The advent of a series of Chiral Stationary Phases (CSP's) for use in High-Performance Liquid Chromatography has opened up the accessibility of that form to direct stereochemical analysis. There are several excellent reviews and classifications of the main types of CSP available (e.g. ref.[3]). As with all separative techniques, the conditions governing resolution will need to be optimised with respect to peak separation, and analysis time. Since for the most optical resolutions there are only two species, and consequently peaks, involved, at first sight optimisation may seem trivial. However it may be dangerous to make assumptions about the relationship between the various parameters affecting resolution, as sequential, single parameter optimisation can lead to the location of a local, rather than the global optimum. Several methods or approaches are available for investigating the entire response surface, as it is described.

OPTIMISATION STRATEGIES

Factor Analysis

The mathematical technique of factor analysis[4] can be used to yield the number of parameters controlling a particular phenomenon. Theoretically therefore, the method could be used to calculate the factor space for controlling resolution on a chiral stationary phase.

Iterative Lattice

Method development by lattice design is a process which allows a chromatographer to identify the important factors controlling the separative process, and also to determine the relative importance of these parameters. Its main application is found in the optimisation of resolution controlled by ratios of two or more parameters, which is usually the case in chiral separations, although quite sophisticated computer programs are required[5].

Response Surface Mapping

Using a Chromatographic Response Function [CRF][5], it is possible to assign a mathematical value to a chromatogram, which takes into account the resolution of all adjacent peaks, and the time taken to achieve that resolution. The higher the arithmetic value of the CRF, the 'better' the system. Using two parameters governing resolution, for example pH and concentration of organic modifier, and plotting CRF against these will result in a three-dimensional response surface, high points on which will represent better chromatographic performance. More usefully, the data can be plotted in the form of a contour diagram, enabling more exact location of high points. This approach has been applied to the resolution on an α_1-acid glycoprotein (Enantiopac) column[6], of the drug oxamniquine[7] (Fig. 1). The system was optimised for pH (over the range 4.50 to 7.00) and percentage of propan-2-ol (from 0.2 to 1.5% v/v), the CRF being used to provide quantitatively useful resolution in as short an analysis time as possible (see Fig. 2.)

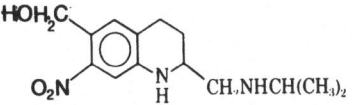

Fig. 1. Molecular structure of oxamniquine.

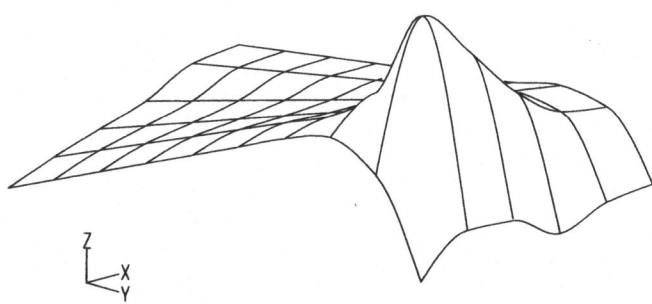

Fig. 2. Response surface for oxamniquine. x = pH; y = Percentage propan-2-ol; z = CRF

Modified Sequential Simplex Method

The modified sequential simplex method (MSSM) has been applied to many chromatographic optimisations[5]. The algorithm selectively directs the chromatographer through the response surface to the point representing optimal conditions. This it does by assigning a CRF to the vertices of a triangle, each point representing a particular set of binary conditions. The resultant trio of chromatograms are studied and designated 'worst', 'next to worst' and 'best', according to their CRF. The worst point is rejected, and a new set of conditions found by reflecting the worst vertex through the mid-point of the line joining the other two points. This is the standard simplex method. The MSSM differs by the addition of the concept of extension and contraction. Basically, if the new point on reflection gives a set of conditions that are worse than those of the original worst point, then the algorithm is 'contracted' - the new point is found by halving (or in fact any other fraction) the line from the vertex in question to the mid point of the line joining the other two points, giving a new set of conditions within the original triangle. Alternatively, if the reflected point yields an eluent giving a much improved CRF, then the line can be extended in that direction by any desired factor (usually a factor of 2). This enhancement of selectively favouring development in the direction of the optimum speeds up the final location of optimal conditions[5]. MSSM has been applied to the resolution of oxamniquine on the AGP columns[7]. The response surface had been shown to contain only a single sharp maximum, therefore the algorithm would not be 'trapped' in finding a 'local optimum'. The method discovered an optimum in 13 steps. This proved to be the same as that obtained from the conventional and more time-consuming study of the response surface 'map' (see Fig. 3).

DEVELOPMENT TRENDS

Novel Phases

At present, although there are in excess of 35 commercially available CSP's, none of these have been shown to be generally applicable. The hope is that in the future, phases will be developed that have a much greater range for handling suitable analytes, although at the time of writing this prospect does appear remote. It would probably be more realistic to expect the development of 'tailor-made' phases for the resolution of particular compounds and closely related species.

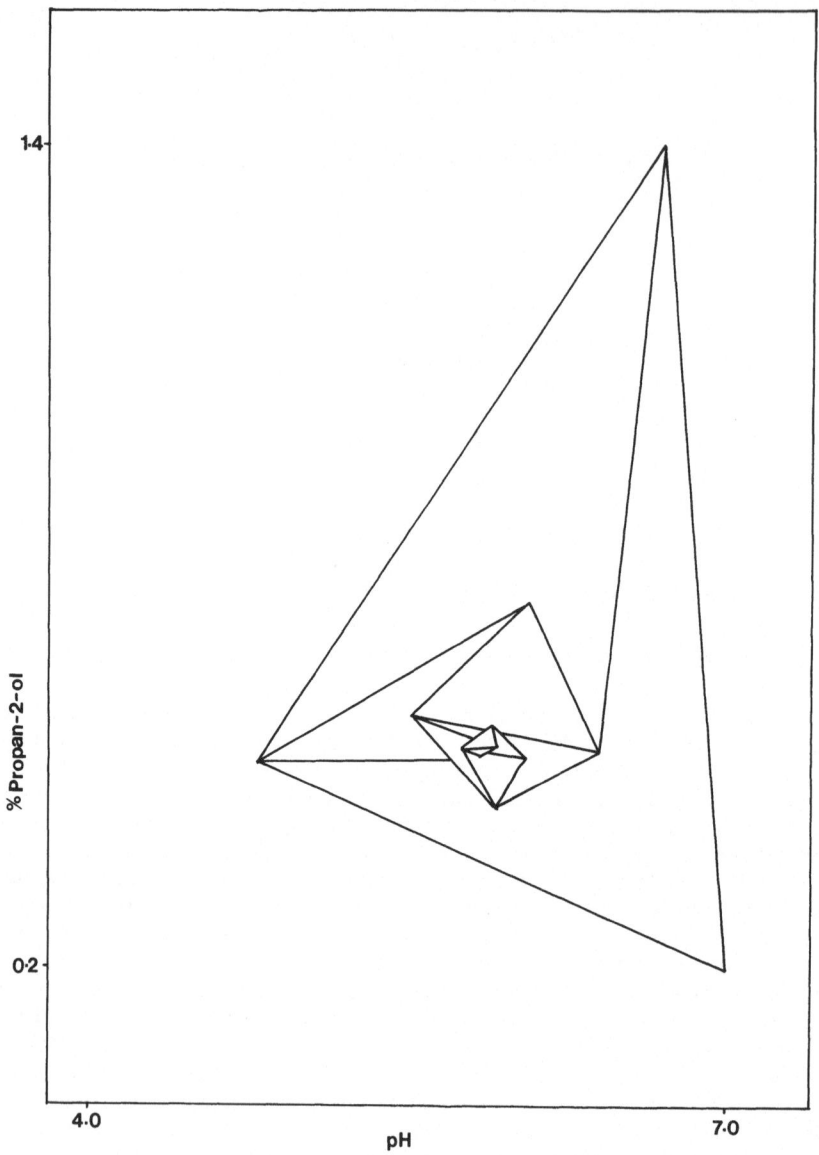

Fig. 3. Modified Sequential Simplex plot for the optimisation of oxamniquine.

Mechanistic Studies

Apart from the 'Pirkle' series of phases, the mechanistics of which were arguably a precursor to the actual phases themselves, and the increasing body of knowledge on cyclodextrin phases, little is understood about the exact mechanism of the chiral recognition process. A greater understanding of the precise series of events associated with differential enantiomeric binding, in terms of the inter-molecular forces (both attractive and repulsive) involved and the steric requirements (for groups immediately adjacent to, as well as removed from, the chiral centre), for a series of CSP's would greatly facilitate the *a priori* selection of the appropriate column. Consequently, greater understanding of the mechanism of resolution for all of the available chiral stationary phases can only assist the analyst wishing to resolve optical isomers in the selection of an appropriate column for a particular problem. This would be preferable to the trial and error procedure, coupled with serendipity, required at present.

124

Chiral Detectors

Enantiomers will behave in exactly the same fashion in all symmetrical environments. For this reason absorption of non-polarised radiation will be similar for both members of an enantiomeric pair. Standard ultra-violet absorption will therefore not distinguish between optical antipodes as they elute from an HPLC column, even if a complete spectrum is obtained, as in the case of linear diode-array detection. The consequence of this is that it is impossible to characterize enantiomers as they elute, unless samples of the pure isomers are available for retention time matching. Usually however this is not the case, and a method is required that will identify the enantiomers as they are detected. Such a method is circular dichroism (CD) detection. This records the difference in absorbance by a species of left and right circularly polarised light. At a maximum in the CD spectrum, this difference will be positive for one enantiomer, and negative for the second. The spectrum itself can be used in the visualization of the interaction of the helical nature of the circularly polarized light, and the groups around an asymmetric atom. This information can then be used to assign an absolute configuration (i.e. (R) or (S)) to the molecule concerned.

Semi-Preparative and Up-Scale Separations

As mentioned earlier, a probable development in the technology of CSP's will be the specific development of a phase to suit a particular analyte. As a consequence of this, the phase will be designed and fine tuned to give extremely high selectivity. This will mean that the column can be effectively grossly overloaded, albeit with large loss of performance, while maintaining adequate separation of the two peaks. This would form the basis for preparative scale enantiomeric separation, perhaps even on a commercial scale. An interesting recent development has been the use of aqueous liquid membranes containing cyclodextrins to selectively transport enantiomers from a bulk racemic solution, as described by Armstrong[8].

Artificial Intelligence and Expert Systems

When the database on separation mechanisms has expanded, the use of expert systems would potentially facilitate selection of column or chiral eluent. This greater understanding of the mechanisms of action of the available chiral stationary phases, in terms of structural requirements for resolution would permit a systematic approach in the method design of optical resolutions to be developed[9].

REFERENCES

1. F. P. A. Lehmann, *Trends Pharmacol.Sci.*, 7, 7:281-85 (1986).
2. P. F. White, J. Ham, W. L. Way, A. J. Trevor, *Anesthesiology,* 52:321-39 (1980).
3. I. Wainer, *Trends Anal.Chem.*, 6,5:125-34 (1987).
4. J. K. Strasters, H. A. H. Billiet, L. De Galan, B. G. M. Vandeginste, G. Kateman, *J.Chromatogr.*, 385 1:181-200 (1987).
5. J. C. Berridge, "Techniques For The Automated Optimisation of HPLC Separations," Wiley, p. 152-64 (1985).
6. J. Hermansson, *J.Chromatogr.*, 298:67-78 (1984).
7. A.F. Fell, T.A.G. Noctor, J.E. Mama, B.J. Clark, J. Chromatogr., 434: (1988) in press.
8. D. W. Armstrong, H. L. Jin, *Anal.Chem.*, 59:2237-41 (1987).
9. T. P. Bridge, M. H. Williams, G. G. R. Seaton, A. F. Fell, *Chromatographia* , 24:691-696 (1987).

A NOTE ON FURTHER USE OF COMPUTER AIDED CHEMISTRY TO PREDICT CHIRAL SEPARATIONS IN LIQUID CHROMATOGRAPHY: SELECTING THE MOST APPROPRIATE DERIVATIVE

Ulf Norinder* and E. Goran Sundholm**

*Computer Graphics Unit and **Pharmaceutical Analysis
Research and Development Laboratories
Astra Alab AB, S-151 85 Sodertalje, Sweden

INTRODUCTION

The formation of transient adsorbates with different stability is essential for the efficient separation of optical isomers on a chiral stationary phase (CSP). To be more precise: The stereoselection occurs only during the time when all binding sites are effectively interacting. A deeper knowledge in these processes is, therefore, a prerequisite for understanding chiral resolution.

In this instance, computer aided chemistry might prove useful as an alternative method to more conventional approaches. Until now, very little has been done in this area. Lipkowitz et al,[1,2] developed techniques to describe the conformational space of a number of well-known Pirkle CSPs. Conformational analysis indicated great conformational flexibility, which in turn might explain their effectiveness for resolving a large number of chiral compounds.

Recently, we described a similar approach to conformational analysis, but we also extended the model to include intermolecular energy calculations[3]. Using that model, we were able to make predictions about quite a large number of different chiral compounds and phases, all with great success. Here we extend the use of the model to test the predictability for separating different amide derivatives of the antidepressant alaproclate (R=H, in structure below) on a Supelcosil® LC-(R) Urea column.

EXPERIMENTAL

Apparatus

The chromatography was performed with a Model 510 pump, a Lambda-Max Model 480 variable wavelength UV detector, a WISP 710B autosampler (all from Waters Associates, Milford, MA) and an SP 4270 integrator (Spectra Physics, San Jose, CA). The column was a 25 cm x 4.6 mm stainless steel Superlcosil® LC-(R)-Urea column (Supelco Inc., Bellefonte, PA). The eluant consisted of hexane: t-butanol: acetonitrile (100:4:1) and was equilibrated in the column overnight. The water content of the eluant was determined by K-F titration to 0.045% (w/w). The flow-rate was 2.0 ml/min, and the detection wavelength 254 nm. Sample solutions (ca 0.04 mg/ml) in hexane were injected (20μl) at ambient temperature (24°C).

3,5-Dinitrobenzoyl chloride, 1-naphthoyl chloride, and 2,3,4,5,6-pentafluorobenzoyl chloride were from Fluka AG, Switzerland.

To ca 1 mg of alaproclate hydrochloride (Astra Alab AB, Sodertalje, Sweden) in a glass tube (4ml) with a stopper, 1 ml of 0.1 M sodium hydroxide was added. Then, 1 ml of a solution of the acid chloride (ca 3 mg/ml in methylene chloride) was added and the mixture thoroughly shaken for 1 min with a Whirlimixer (Fisons Scientific Apparatus, England). After separation, the organic layer was washed with pH 2 phosphate buffer and filtered through a pipette fitted with a cotton plug and ca 50mg magnesium sulphate. Mass spectra (LKB 9000, Bromma, Sweden, direct inlet) agreed with the expected structures, which are shown below.

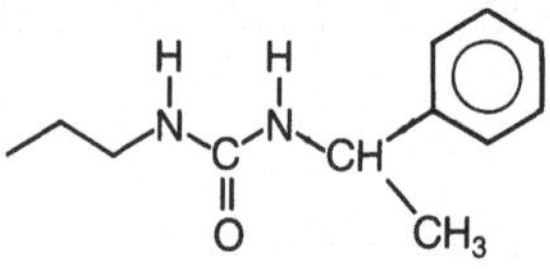

1

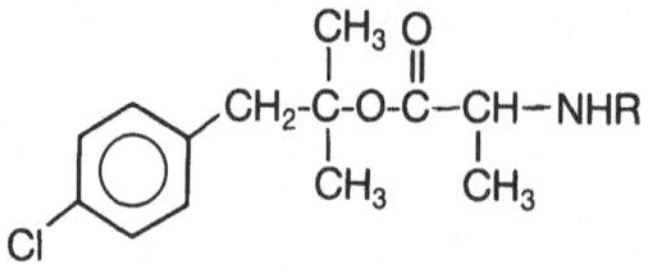

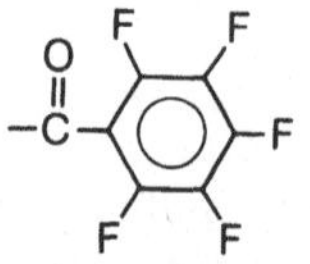

Method of Calculation

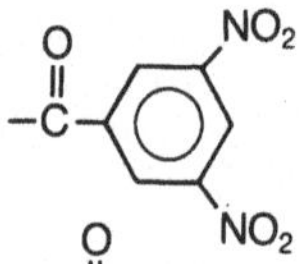

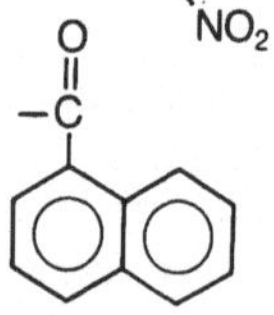

Details of the method used for calculating intermolecular interaction energies as well as orientations have been described elsewhere[3].

RESULTS AND DISCUSSIONS

The separation of the (+)/(-) dinitrobenzamides of alaproclate (2) on the LC-(R)-Urea column is shown in the Figure. Interaction energies calculated by the model and chromatographic parameters are presented in the Table. Compound 1 was used as a model for the CSP. Also, the differences in free energies $-\Delta(\Delta G°)$, calculated from α-values according to the equation

$$-\Delta(\Delta G°) = RT \ln \alpha,$$

have been included.

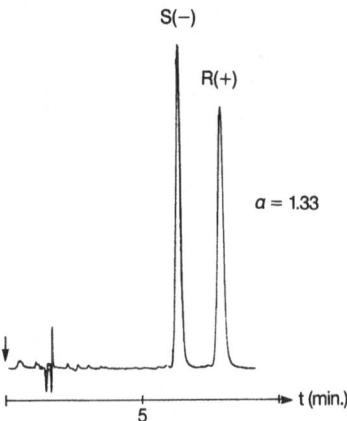

Fig. 1.   Enantiomeric resolution of the DNB amides of alaproclate (2) on a Supelcosil® LC-(R)-Urea column.

Table   Interaction Energies and Chromatographic data of Compounds 2-4

Compound No.	Calculated energy (kcal/mol) R-form	S-form	Energy diff. (kcal/mol)	RT ln α (kcal/mol)	k'₁	α	Con-figuration[a]
2	-17.86	-17.35	0.51	0.17	3.0	1.33	R
3	-13.88	-13.78	0.10	0.03	1.9	1.06	S
4	-16.67	-16.59	0.08	0	0.5	-	-

[a] Absolute configuration of the second eluted enantiomer

The eluent was optimized such that reasonable retention times could be achieved. The addition of acetonitrile had little effect on the separation selectivity. However, peak symmetry was considerably improved by a small amount of this modifier. The pentafluorobenzamides (4) could not be resolved in any eluant chosen.

The R-form of 2 interacts more strongly with the CSP, which is in agreement with calculated data. With the 1-naphtamides (3) however, the R-form is eluted first (Table), although calculations predict the opposite. The energy difference of the adsorbates for such a small separation (α = 1.06) is obviously too small to yield reliable results in this case. This is not surprising keeping in mind the approximations necessary in the use of computer aided chemistry[3]. The magnitude of this small energy difference is of the same order as is obtained in the case of the pentaflourobenzamides (4) where no separation at all is observed. Despite this short-coming, the model is still very useful. It predicts which of the amides is most likely to be separated. This knowledge is not, a priori, easily obtainable from the use of hand-held models or from simple bonding notations. Further refinement of the model as well as the development of more powerful computational techniques might improve predictability and aid in designing new chiral stationary phases for chromatographic purposes.

REFERENCES

1. K. B. Lipkowitz, D. J. Malik and T. Darden, *Tetrahedron Lett.* 27:1759 (1986).
2. K. B. Lipkowitz, D. J. Malik and T. Darden, *Anal.Chem.* 58:1611 (1986).
3. U. Norinder and E. G. Sundholm, *J.Liq.Chromatogr.* 10:2825 (1987).

A *NOTE ON* AN OPTICAL ROTATION DETECTOR FOR HIGH-PERFORMANCE

LIQUID CHROMATOGRAPHY

D.M. Goodall and D. K. Lloyd

Chemistry Department
University of York
Heslington, York YO1 5DD, UK

INTRODUCTION

There have been several examples of enantiomeric purity determination without separation or with incomplete separation[1,2]. Generally the usefulness of this technique has been limited by the lack of sensitivity of commercially available polarimeters. Recently Reitsma and Yeung have demonstrated enantiomeric purity determinations of submicrogram quantities of material[3] using an Argon ion based laser polarimeter[4,5].

The optical-rotation (OR) HPLC detector described here has been developed from work on a stopped flow polarimeter for fast reaction studies[6,7]. Using a compact semiconductor laser light source and a photodiode detector it has sufficient sensitivity to detect microgram injections of optically active material.

INSTRUMENT DESIGN

The polarimeter uses the Faraday modulation polarimetry technique whereby the polarisation direction of light passing through the system is modulated at some frequency f_{mod}[6]. This produces an amplitude modulated signal at the photodetector with a signal proportional to the sample rotation at f_{mod}, an unwanted carrier component at 2 f_{mod} and a dc component due to the imperfect extinction ratio of the polariser/analyser. The sample signal at f_{mod} is recovered using a phase sensitive detector. A block diagram of the detector is shown in Fig. 1.

The light source is a collimated laser diode (Mullard CQL13A) giving up to 2 mW of light at 820 nm. The polariser and analyser are Glan Taylor prisms (Rofin G15). Polarisation modulation is provided by a Faraday modulator consisting of 1500 turns copper wire around a 1cm diameter, 5 cm long dense flint glass rod. Various cells have been used with linear or tapered bores and flat or lensed end windows, with volumes from 1 μl to 20 μl. The photodetector is a silicon photodiode (Centronics OSD15-5) with a low noise preamplifier. A phase sensitive detector is used for signal recovery.

Detector performance may be limited by shot or flicker noise, which are dependant on the optical and mechanical properties of the system, or by noise in the detector electronics[8]. The semiconductor diode laser used has very low amplitude noise, measured to be 0.0075% rms in the band 10 Hz to 14 kHz. This allows a high light intensity at the detector and thus a large modulation angle to achieve a good signal to noise (S/N) ratio with only ordinary commercial prism polarisers. Our present detector is limited by noise in the photodiode amplifier, giving a short term noise level of 4 μ^orms(3s detector time constant).

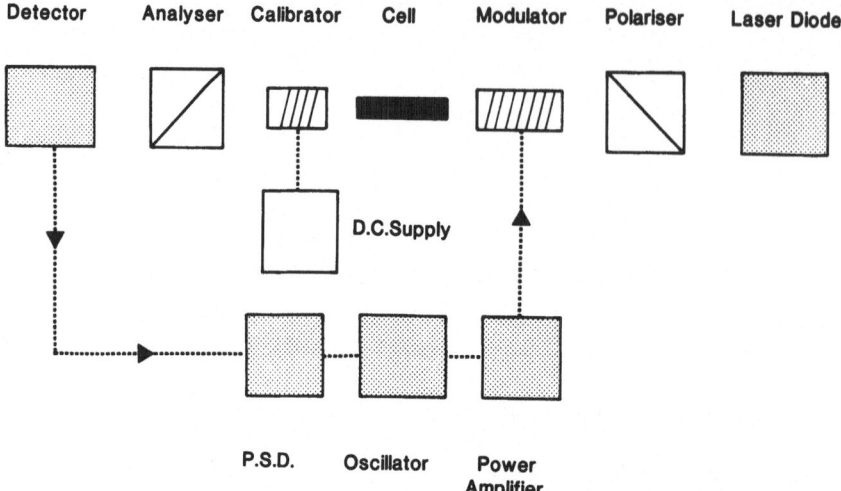

Detector   Analyser   Calibrator   Cell   Modulator   Polariser   Laser Diode

D.C.Supply

P.S.D.   Oscillator   Power Amplifier

Fig. 1. Optical rotation detector block diagram.

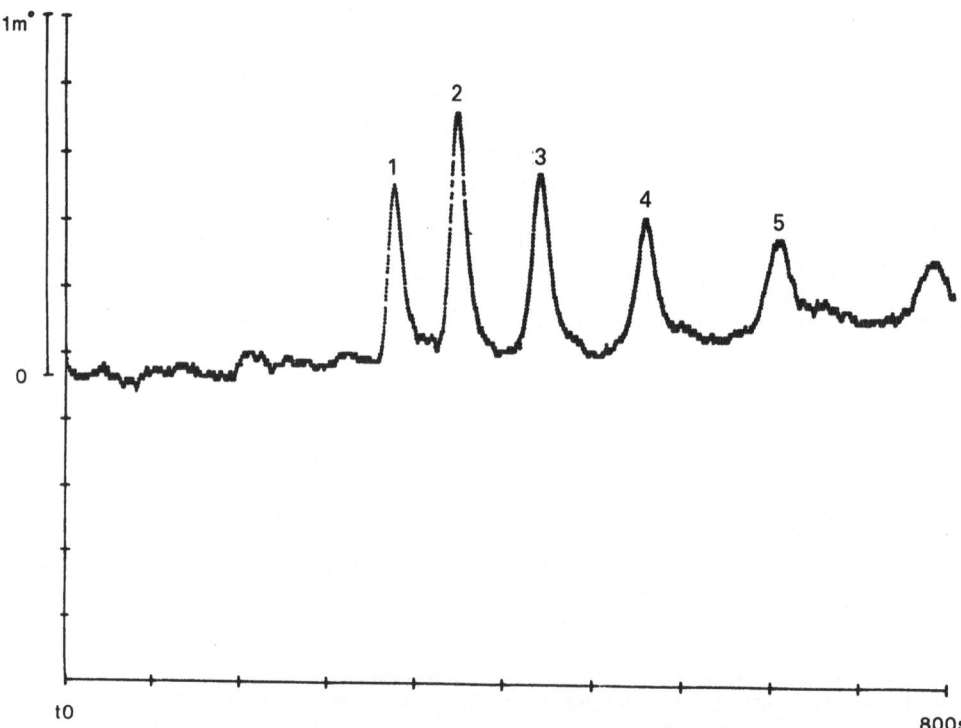

Fig. 2.   Chromatogram of glucose syrup with optical rotation detection.   Column: Spherisorb S5, amino bonded; Solvent: acetonitrile - water 65:35 (v/v); Flow rate: 1 ml/min; Cell: Applied Chromatography Systems, 1 cm pathlength; Laser: 4mW GALA diode, 750 nm; Sample: 10 microlitre glucose syrup, 440 microgram total sugars; Peaks: 1. glucose, 80 µg; 2. maltose, 67 µg, 3. maltotriose, 58 µg, 4. maltotetrose, 5. maltopentose.

Fluctuations over tens of seconds or minutes are due to mechanical instabilities in the optical bench layout. When a 20 mW version of the Mullard diode becomes available an improvement in S/N should be obtained, with shot and flicker noise being the limiting factors. Other collimated diodes have been used but generally beam quality is not as good, making it more difficult to focus the beam through the cell.

APPLICATIONS

The separation of glucose syrup detected using our dveice is shown in Fig. 2. A Spherisorb S5 amino bonded column with acetonitrile: water (65:35 v/v) as solvent at a flow rate of 1 ml/min-1 was used. For this chromatogram a 4 mW GALA laser diode system was used with a 1cm pathlength 8 μl volume cell. The second peak, due to maltose, has a larger area than the glucose peak despite the smaller quantity injected because of its higher specific rotation.

REFERENCES

1.   A. Mannschreck, M. Mintas, G. Becher and G. Stuhler, *Angew.Chem.Int.Ed.Engl.*, 19:353 (1980).
2.   W. Boehme, G. Wagner and U. Oehme, *Anal.Chem.* 54:709 (1982).
3.   B. H. Reitsma and E. S. Yeung, *J.Chromatogr.*, 362:353 (1986).
4.   E. S. Yeung, L. E. Steenhoek, S. D. Woodruff and J. C. Kuo, *Anal.Chem.* 52:1399 (1980).
5.   D. R. Bobbitt and E. S. Yeung, *Appl.Spectrosc.*, 40:407 (1986).
6.   D. M. Goodall and M. T. Cross, *J.Sci.Instrum.*, 46:391 (1975).
7.   D. M. Goodall and D. K. Lloyd, *Gums and Stabilisers for the Food Industry 3*, Elsevier, London, p. 497 (1986).
8.   E. S. Yeung, *Talanta*, 32:1097 (1985).

PROSPECTS FOR CHIRAL THIN-LAYER CHROMATOGRAPHY

I.D. Wilson and R.J. Ruane

Department of Safety of Medicines
ICI Pharmaceuticals, Mereside, Alderley Park
Macclesfield, Cheshire SK10 4TG, UK

SUMMARY

The potential benefits arising from the development of robust and efficient methods for enantiomer separation by thin-layer chromatography (TLC) are discussed and the current state of the art in chiral TLC reviewed. The relative lack of progress in developing suitable chiral stationary phases (CSP s) for TLC of wide general applicability is noted and contrasted with advances in chiral methodology for high performance liquid and gas liquid chromatography. The need for further research into the preparation of CSP s, to enable enantioselective separations to be performed on TLC is highlighted and recent progress towards this goal described.

INTRODUCTION

Despite a decade or more of rapid improvements in the type, quality and performance of the plates used for thin-layer chromatography (TLC) and the dramatic advances in the instrumentalization of the technique TLC still suffers from a poor image. Thus, the perception is widespread amongst chromatographers that TLC is obsolete, outmoded and incapable of reproducible, high resolution, separations and therefore quite unsuited to the demands of modern analytical chemistry. Indeed, the general feeling of many analytical chemists was perfectly captured at this symposium by the remark (by one of the most eminent workers in the field of chiral separations) "Well who would want to do chiral separations on TLC anyway?" This attitude probably reflects the fact that most analysts' experiences of TLC predate the development of high performance (HP), high quality, TLC plates, and the availability of rapid and versatile scanning densitometers. However, in the same way that high performance liquid chromatography (HPLC) evolved from open glass columns, packed with large and irregular particles of silica or alumina, with solvents fed by gravity and the eluate collected in fractions for subsequent analysis by other methods, so TLC has also developed from its similarly unpromising beginnings into a modern instrumental technique. This article was written in response to the short debate which occurred at the symposium as to the relative merits of "chiral TLC" versus other modes of chromatography, and in it we briefly survey the current areas of research in this area.

Of course all of these developments would be irrelevant if, in some circumstances, TLC did not offer positive advantages over other techniques. Therefore, before embarking on a discussion of chiral TLC, the strengths and weaknesses of TLC as an analytical technique ought to be restated.

A major advantage of TLC over many other chromatographic techniques is its flexibility. For example, in TLC, detection and evaluation is generally an "off line" procedure. This allows a visual evaluation of the plate before scanning densitometry etc., and those places, or individual tracks, which are of particular interest can be dealt with first. Similarly any plate, or track, which does not meet the required criteria need not be subjected to further analysis. Imagine the benefits which would accrue for HPLC or GLC if samples could be rejected prior to detection, but after chromatographic separation! In addition, because chromatography and detection are not directly linked, many plates for different assays can be run at the same time and scanned when time allows, the detector is not committed to a particular assay at a particular time. Also chromatography may be performed at a site some distance from the detector, at the point of sampling for example, before being taken to the central laboratory for quantification. This property of the TLC plate, whereby it can be used to store the chromatogram, also allows great flexibility in identification. Thus following scanning densitometry, the components of interest on the plate can be further analyzed spectroscopically and *in situ* UV, IR and mass spectra obtained. More simply, any of the many specific (or non specific) color forming reactions which have been devised for the detection of particular compounds or classes on TLC plates can be used (analogous to post-column reaction in HPLC).

TLC is also useful in those situations where a high throughput of samples is required, because even where run time is slow, the fact that many samples (up to 60 samples on a 10 x 20 cm HPTLC plate) are being "parallel processed" effectively reduces the total time per sample to a few minutes.

The technique is cheap, robust and of course not subject to the same limitations in terms of solvent that govern the choice of mobile phase composition in HPLC. This means that not only can UV opaque solvents be used but also that aggressive solvents (e.g. high pH etc.) and samples, which would rapidly degrade expensive HPLC columns, can be chromatographed without difficulty.

A further advantage of TLC is, of course, that, providing the components in the sample are not volatile then they will be located on the plate somewhere between the origin and the solvent front. Unlike HPLC therefore there is no danger that the material present in the sample will go undetected due to retention on the column.

These therefore, are some of the strengths of TLC. The major weaknesses of TLC include the relatively low resolving power available on the plate compared to HPLC or GLC and the fact that limits of detection are sometimes higher for TLC than for either HPLC or GLC. It is also rather more difficult to fully automate TLC to the extent that has proved possible for HPLC and GLC although important steps have been made in this direction.

ENANTIOMER SEPARATION ON TLC

For TLC, as for the other major chromatographic techniques, two different approaches to enantiomer resolution can be readily envisaged. In the first instance the compounds of interest may be converted into diastereoisomers by derivatization with a suitable chiral derivatizing reagent. The separation of diastereoisomers by TLC is well established, and certainly no more difficult than by any other technique, and numerous examples of the successful application of this approach exist in the literature. However, the formation of diastereoisomers by derivatization, whilst superficially attractive, is not without difficulties. As has been pointed out on many occasions such procedures depend on the availability of reagents of high enantiomeric purity, require good yields without the possibility of stereoselectivity in the rate of formation of derivatives, and must not allow changes in the stereochemistry of the analyte to occur on reaction. For these reasons, which are clearly of general applicability and not specific to TLC, the use of chiral chromatographic systems, either solvents or adsorbents, which enable the direct resolution of enantiomers (or achiral derivatives thereof) are to be preferred. It is this aspect of chiral separations on TLC which will be considered here.

RESOLUTION OF PHENYLTHIOHYDANTOIN AMINO ACID ENANTIOMERS ON (+)-TARTARIC ACID IMPREGNATED SILICA GEL

Recently Bhushan and Ali[1] have reported the successful resolution of a variety of phenylthiohydantoin (PTH) derivatives of a number of D and L amino acids. The system employed for this work was based on the impregnation of silica gel with a pure, optically active, acid (e.g. tartaric acid, ascorbic acid or alanine). For (+)-tartaric acid a slurry of silica gel (50g) in water (100 ml), containing 0.3g of the acid, was made up and TLC plates prepared from it. Following heating at 60°C for 6 to 8 hours the plates were allowed to cool to room temperature and the samples applied in ethyl acetate. At this point the plates were heated again at 60°C for 10 minutes before being developed in chloroform-ethyl acetate-water (28:1:1 V/V/V) for 35 minutes. After drying the PTH amino acid enantiomers were visualized in an iodine chamber. As shown in Table 1 excellent separations were obtained for a total of 9 DL amino acid mixtures. The Rf values obtained for the L-PTH-amino acid enantiomers were identical to those of pure standards.

The attempts to use (+) ascorbic acid and (-)-alanine were less successful than for (+) tartaric acid. However, using the solvent system n-butylacetate-chloroform (1:5 V/V) it proved possible to resolve the PTH derivatives of DL-methionine phenylalanine, valine, threonine, alanine and serine using both (+)-tartaric acid and (+) ascorbic acid for impregnation of the silica.

This methodology is clearly cheap, rapid and easy to use for the resolution of this type of amino acid derivative. Whether it has more general applicability has yet to be determined.

RESOLUTION OF AMINO ACID ENANTIOMERS ON CELLULOSE

In an early example of the use of a chiral stationary phase for TLC, crystalline cellulose (Avicel, E. Merck), Yuasa et al.[2] investigated the separation of DL-tryptophan and a number of other DL-amino acid mixtures. Separations were performed on 10 x 10, and 20 x 20 cm plates in a variety of solvent systems. Although the paper is not particularly clear on this point the solvent systems used would appear to have been based on sodium citrate buffer (0.2M, pH 6.4)-ethanol (1:1 V/V) containing either pyrazine, phenol, imidazole, pyrazole or pyridine-water-alcohol (either ethanol, propanol or butanol) at 5 g/l. According to the authors all the systems investigated yielded "excellent separations" of D and L tryptophan, but systems containing either pyridine or imidazole were selected for the most detailed study. The best results appear to have been obtained with the solvent system ethanol-pyridine-water (1:1:1 V/V/V) which enabled the resolution of DL-tryptophan, DL-histidine, DL-phenylalanine, DL-tyrosine and DL-B-3,4-dihydroxyphenylalanine (but not DL-threonine) using ascending chromatography for 6 hours at 25°C. With the exception of histidine the D amino acids had higher Rf values than the L.

SEPARATION OF ENANTIOMERS ON β-CYCLODEXTRIN BONDED TLC PLATES

In the same way that β-cyclodextrin bonded phases have been used for HPLC, some limited applications have been reported for TLC[3,4]. The β-cyclodextrin bonded phases were produced, using published procedures, for a variety of different manufacturers' silica gels, and in addition a number of different TLC binders were used to prepare the plates. The plates themselves (5 x 20 cm, glass backed) were prepared by mixing 1.5g of the bonded phase and 0.002g of binder together in 15 ml of aqueous methanol (1:1 V/V). This was then spread on to the plate to give a layer approximately 3 mm thick. These plates were then heated to 75°C for 15 minutes prior to use, resulting in a final layer thickness of less than 1 mm. Thus prepared, these plates were then used to separate a variety of optical, positional and geometrical isomers using a mobile phase of methanol-water (1:1 V/V) containing 1% aqueous triethylammonium acetate (pH 4.1). During the course of this work it was found that a silica gel produced by Macherey-Nagel & Co. of between 5 to 20 μm particle size and with 60 Å pore diameter, gave the best results. Experiments with the different types of binder also showed that careful choice was essential for good results. Overall a polymeric binder, "ASTEC all

Table 1. hR$_f$ of PTH-amino acid enantiomers[1]

DL mixtures of PTH-amino acids	hR$_f$ *	
	D	L
Methionine	16	83
Phenylalanine	15	85
Tryptophan	-	95
Valine	21	80
Isoleucine	15	92
Tyrosine	16	95
Threonine	30	85
Alanine	12	55
Serine	10	84

* hR$_f$ = R$_f$ x 100
Solvent: chloroform-ethyl acetate-water (28:1:1 V/V/V).
Development time: 35 min. Solvent front: 10 cm. Room temperature: 25 ± 1°C.

Table 2. hR$_f$ values for various enantiomeric compounds on β-cyclodextrin bonded TLC plates[3]

Compound	hR$_f$	Mobile phase (V/V)[b]
dansyl-D-leucine	49	40/60
dansyl-L-leucine	66	
dansyl-D-methionine	28	25/75
dansyl-L-methionine	43	
dansyl-D-alanine	25	25/75
dansyl-L-alanine	33	
dansyl-D-valine	31	25/75
dansyl-L-valine	42	
D-alanine-β-napthylamide	16	30/70
L-alanine-β-napthylamide	25	
D-methionine-β-napthylamide	16	30/70
L-methionine-β-napthylamide	24	
(-)-1-ferrocenyl-1-methoxyethane	31	90/10
(+)-1-ferrocenyl-1-methoxyethane	42	
(-)-1-ferrocenyl-2-methylpropanol	33	90/10
(+)-1-ferrocenyl-2-methylpropanol	39	
(-)-S-(1-ferrocenylethyl)thioglycolic acid	37	90/10
(+)-S-(1-ferrocenylethyl)thioglycolic acid	44	

β-cyclodextrin bonded to Macherey-Nagel silica gel with ASTEC "all solvent binder".
[b] ratio of methanol to 1% triethylammonium acetate (pH 4.1). (+) or (-) refers to the Cotton effect of 250 nm not the optical rotation at the sodium D line.

solvent binder", produced the best spot shape etc. Using β-cyclodextrin bonded to the Macherey-Nagel silica gel a range of dansyl amino acids and other enantiomers were well resolved (see Table 2).

The D enantiomer of all the dansyl amino acids had the highest R$_f$ value with the separation between D and L-dansylmethionine being highlighted as particularly impressive.

This type of TLC plate clearly represents a relatively simple development of current chiral phases for HPLC, and could presumably be commercialized relatively easily.

Given that both mobile phase and elution behavior were claimed to be similar to those obtained by HPLC such TLC plates could have a valuable role in screening prior to the use of such a phase for HPLC.

SEPARATION OF 2,2,2-TRIFLUORO-1-(9-ANTHRYL) ETHANOL ON IONICALLY MODIFIED (R)-N-(3,5-DINITROBENZOYL) PHENYLGLYCINE AMINOPROPYL TLC PLATES

The separation of enantiomers of 2,2,2-trifluoro-1-((-anthryl) ethanol using a chiral stationary phase on TLC was achieved by Wainer et al. [5]. In this work γ–aminopropyl bonded silica gel (Zorbax BP-NH$_2$) was coated onto microscope slides and then converted to a "Pirkle" type CSP by ionically bonding (R)-N-(3,5-dinitrobenzoyl) phenylglycine to the bonded silica. This was achieved by continuous development of the plates in a tetrahydrofuran solution of chiral modifier (1g/20 ml). The degree and extent of coating could be monitored visually as the resulting CSP was pale yellow in color. After coating, the plates were air dried, and an aliquot of racemic 2,2,2-trifluoro-1-(9-anthryl) ethanol applied (5 μg). The plates were then developed using a solvent system consisting of hexane-isopropanol (9.5:1 V/V) resulting in the appearance of two fluorescent spots with R$_f$ values of 0.59 and 0.49 for the (-) and (+)-isomers respectively.

Since this report aminopropyl bonded TLC plates have become commercially available, greatly simplifying the task of preparing such plates. Care is, however, required in their use as our own experience of this type of CSP is that it can rapidly discolor (probably due to oxidation) and is then useless for chromatography[6].

SEPARATION OF 2,2,2-TRIFLUORO-1-(9-ANTHRYL)-ETHANOL ON IONICALLY MODIFIED (L)-N-(3,5-DINITROBENZOYL) LEUCINE AMINOPROPYL TLC PLATES

In our own (unpublished) studies we have investigated the use of both the (R)-N-(3,5-dinitrobenzoyl) phenylglycine and (L)-N-(3,5-dinitrobenzoyl) leucine ionically bonded to commercial aminopropyl TLC plates (E. Merck, HPTLC 10 x 10 cm cat. no. 15647). In these studies the latter CSP gave better resolution of 2,2,2-trifluoro-1-(9-anthryl) ethanol using a solvent system of hexane-isopropanol (9:1) and plates coated (by dipping) in a 0.05M solution of the reagent in tetrahydrofuran. The concentration of reagent used to coat the plates does seem to be important as the results obtained with 0.001 and 0.01M coating solutions were inferior to that obtained using 0.05M solution. The discoloration of both TLC plates and reagent solutions appeared to be minimized by storage in darkness.

RESOLUTION OF ENANTIOMERIC AMIDES ON (R)-(-)-1-(NAPTHYL)-ETHYLISOCYANATE COVALENTLY BONDED TO AMINOPROPYL TLC PLATES

Recently Brunner and Wainer[7] have reported the use of a CSP originally developed by Oi et al. [8] for the resolution of enantiomeric amides of a variety of compounds, including amides produced from ibuprofen.

Plates were prepared by soaking a commercial aminopropyl bonded silica gel TLC plate (Alltech) in 20 ml of a dichloromethane solution of (R)-(-)-1-(1-napthyl)-ethylisocyanate (1g/100 ml) for 5 min. Plates were removed from the solution and then air dried, followed by washing twice in dichloromethane. Analytes were converted to amides (3,5-dinitroanilines for acids and 3,5-dinitrobenzoyl derivative for amines). Chromatography was then performed with a solvent system consisting of hexane-2-propanol-acetonitrile (20:8:1 V/V). Good resolution was achieved for a total of 7-enantiomer pairs (5 acids and 2 amines) although the results obtained depended on the derivative formed.

Given the ready availability of both the reagent used to form the CSP and the aminopropyl bonded TLC plates then commercialization of this type of CSP for TLC ought to be relatively straight forward.

RESOLUTION OF AMINO ACID ENANTIOMERS ON TLC USING LIGAND EXCHANGE

Perhaps the one area of chiral chromatography for which a TLC mode has been extensively researched is that involving ligand exchange. Currently this is the only area for which a commercial product (Macherey Nagel's ' "Chiral plate") has been developed. This type of plate was first described, by Gunther *et al.*[9] in 1984 where the separation of a variety of amino acids using an impregnated C-18 bonded TLC plate was discussed. The plate was prepared by first immersing in a 0.25% solution of copper[II]acetate (1 min. methanol-water, 1:9 V/V), followed by drying and then a second immersion in a 0.8% methanolic solution of the chiral selector (1 min). The chiral selector chosen by these authors was (2S,4R,2'RS)-4-hydroxy-1-(2'hydroxydodecyl)-proline. Following drying the plates were used to separate the amino acid enantiomers of isoleucine, phenylalanine, tyrosine, tryptophan, proline, glutamine, and 3-thiazolidine-4-carboxylic acid using either methanol-water-acetonitrile 50:50:200 (V/V/V) or methanol-water-acetonitrile 50:50:30 (V/V/V). Detection was by reaction with ninhydrin (0.1%). A development distance of 14 cm was used and good separations of the various enantiomers was achieved (Table 3). The authors claimed that using such plates the resolution was good enough to detect >1% contamination of one enantiomer with the other.

In a second paper[10] these authors greatly extended the range of racemates separated (Table 4) including amongst them natural and "non-natural" amino acids, N-methylated amino acids, N-formyl amino acids, dipeptides, a lactone derivative and a variety of other amino acid derivatives. In some cases as little as >0.25% of one enantiomer could be detected in the presence of the other.

Development times on these plates were between 30 and 90 minutes. In a later study Brinkman and Kamminga re-investigated the mobile phase composition required to obtain separations of enantiomers[11]. These authors were able to reduce analysis time to 4 minutes using acetonitrile-water 80:20 (V/V) with a 5 cm run.

A number of applications using these ligand exchange plates have now been described[12-17]. Their use has recently been described in combination with scanning densitometry for the control of optical purity of D and L-tyrosine[18]. In this application as little as 0.1% of one enantiomer was detectable in the presence of the other with visualization using ninhydrin and scanning densitometry at 520 nm.

Table 3. Resolution of compounds by ligand exchange[9]

Compound	hR$_f$ (Configuration)		Solvent
Glutamine	37(S)	53(R)	1
Isoleucine	37(2R,3R)	44(2S,3S)	1
Phenylalanine	38(R)	45(S)	1
Proline	40(R)	59(S)	2
3-Thiazolidine-4-carboxylic acid	42(S)	52(R)	1
Tryptophan	39(R)	45(S)	1
Tyrosine	26(S)	34(R)	2

Development for 14 cm Saturated chamber.
1:methanol-water-acetonitrile = 50/50/200 (V/V/V)
2:methanol-water-acetonitrile = 50/50/30 (V/V/V)

Table 4. Resolution of compounds by ligand exchange[10]

Compound	hR$_f$ (Configuration)		Solvent
allo-isoleucine	51(D)	61(L)	1
2-Aminobutyric acid	48	52	1
3-Amino-3,5,5-trimethyl-butyrolactone-HCl	50	59	1
4-Aminophenylalanine	33	47	1
0-Benzylserine	54(D)	65(L)	1
0-Benzyltyrosine	48(D)	64(L)	1
4-Bromophenylalanine	44	58	1
5-Bromotryptophan	46	58	1
3-Chloralanine	57	64	1
S-(2-Chlorobenzyl)-cysteine	45	58	1
4-Chlorophenylalanine	46	59	1
3-Cyclopentylalanine	46	56	1
2-Fluorophenylalanine	55	61	1
3-Fluorotyrosine	64	71	1
N-Formyl-tert.-leucine	48(+)	61(-)	1
N-Glycylphenylalanine	58(L)	57(D)	2
Homophenylalanine	49(D)	58(L)	1
cis-4-Hydroxyproline	41(L)	59(D)	1
4-Iodophenylalanine	45(D)	61(L)	1
Methionine	54(D)	59(L)	1
4-Methoxyphenylalanine	52	64	1
5-Methoxytryptophan	55	66	1
2-(1-Methylcyclopropyl)-glycine	49	57	1
4-Methyltryptophan	50	58	1
5-Methyltryptophan	52	63	1
6-Methyltryptophan	52	64	1
7-Methyltryptophan	51	64	1
N-Methylphenylalanine	50(D)	61(L)	1
4-Nitrophenylalanine	52	61	1
Norleucine	53(D)	62(L)	1
Phenylglycine	57	67	1
S-(3-Thiabutyl)-cysteine	53	64	1
S-(2-Thiapropyl)-cysteine	53	64	1
Valine	54(D)	62(L)	1

Development for 13 cm
1:methanol-water-acetonitrile = 50/50/200 (V/V/V)
2:methanol-water-acetonitrile = 50/50/30 (V/V/V)

In an alternative approach employing ligand exchange Weinstein[19] described a system, also based on the impregnation of C18 bonded TLC plates (Merck 10 x 20 cm or Whatman 5 x 20 cm), for the resolution of dansylated amino acids. The plates were prepared by predeveloping in 0.3M sodium acetate in 40% acetonitrile and 60% water at pH 7 (with acetic acid). After fan drying these plates were dipped into an 8 mM solution of N,N-di-n-propyl-L-alanine and 4 mM copper[II] acetate (in acetonitrile-water 97.5:2.5 V/V) for at least 1 hour "and up to overnight". Alternatively plates were sprayed with this solution. Once treated in this way the plates were left to dry.

Dansyl amino acids, in aqueous solution, were then applied to the impregnated plates which were subsequently developed using the 0.3M sodium acetate-acetonitrile-water mixture used for the first impregnation step (with or without N,N-di-n-propyl-L-alanine and copper[II] acetate (1 mM) dissolved in it.) Following chromatography the dansylated amino acids were detected by irradiation with UV light (360 nm). These plates were claimed to be

141

able to separate all the dansylated protein amino acids, except proline, into their enantiomers. In this system the L-enantiomers had the lowest R_f values. It was noted that variation in the mobile phase was possible depending on the enantiomers to be resolved with 25% acetonitrile preferred for glutamine and aspartic acid. In a subsequent paper Grinberg and Weinstein[20] published a more refined procedure employing a two dimensional gradient, RP-TLC system (C18 bonded TLC plates, E. Merck, 10 x 20 cm) to enable the resolution of complex mixtures of the type frequently encountered in biological samples. The first dimension employed an achiral RP-TLC separation of the dansylated amino acids and used a convex gradient beginning with 0.3M sodium acetate in water-acetonitrile (80:20 V/V, pH 6.3) to which was added 0.3M sodium acetate in water-acetonitrile (70:30 V/V pH 6.8) at a flow rate of 0.5 ml/min to give a final acetonitrile content of 38 or 47%. Following the separation of the various dansylated amino acids by chromatography in the first dimension the plates were dried and the strip containing the amino acids was covered with a glass plate. The remainder of the plate was then sprayed with a solution of N,N-di-n-propyl-L-alanine (8 mM) and copper (II) acetate (4 mM) in acetonitrile-water (95:5 V/V).

In the second dimension the chiral separation was achieved using a mobile phase of 8 mM N,N-di-n-propyl-L-alanine and 4 mM copper (II) acetate in 0.3M sodium acetate in water-acetonitrile (70:30 V/V, pH 7). Development in the second dimension was performed with a temperature gradient of 6.25°C/cm to improve the separation.

This combination of an achiral solvent gradient separation to separate the amino acids into single, albeit racemic, compounds followed by a chiral separation, with a temperature gradient, to resolve the enantiomers clearly represents an elegant and powerful approach. However, the authors noted that development in the second dimension took some $2^1/2$ hours.

Although not of general applicability the separation of amino acids and related compounds using ligand exchange is both easy to implement and gives good separations. Combined with scanning densitometry for quantification it may well provide a viable alternative to HPLC for certain applications.

CONCLUSIONS

This brief survey of chiral TLC demonstrates that, where attempts have been made to develop chiral stationary phases for TLC they have often been surprisingly successful. There have however, by comparison with HPLC, been relatively few studies in this area and this more than anything else is the reason for the poor state of chiral TLC at present. With more effort there is no reason to suppose that very successful TLC-CSP s could be obtained. It would, of course, be absurd to suggest that such TLC-CSP s would, or should, replace those developed for GLC and HPLC, and that is not our intention. However, TLC offers a different spectrum of possibilities compared to these other chromatographic techniques and is likely to remain a major chromatographic technique for the foreseeable future. As recently observed by Friedrich Giess "the strengths of column liquid chromatography are well known, as witnessed by its prodigious popularity, but its weaknesses are often tolerated in the belief that no better alternatives exist. However, TLC represents a totally different approach and avoids many problems inherent in CLC"[21].

More research into CSP s for TLC is clearly necessary but if this is done the future prospects for chiral TLC look promising.

REFERENCES

1. R. Bhushan and I,. Ali, *J.Chromatogr.*, 392:460 (1987).
2. S. Yuasa, A. Shimada, K. Kameyama, M. Yasui and K. Adzuma, *J.Chrom.Sci.*, 18:311 (1980).
3. A. Alak and D. W. Armstrong, *Anal Chem.*, 58:582 (1986).
4. T. J. Ward and D. W. Armstrong, *J.Liq.Chromatogr.*, 9:407 (1986).
5. I. W. Wainer, C. A. Brunner and T. D. Doyle, *J.Chromatogr.*, 9:154 (1986).

6. I. D. Wilson *in* "Bioactive Analytes, including CNS Drugs, peptides and enantiomers", ed. E. Reid, B. Scales and I.D. Wilson, Plenum, 1986.

7. C. A. Brunner and I. W. Wainer, *J.Chromatogr.*, in press.

8. N. Oi, H. Kitahara, T. Doi and S. Yamamoto, *Bunseki Kagaku*, 32:345 (1983).

9. K. Gunther, J. Martens and M. Schickedanz, *Angew.Chem.Int.Ed.Engl.*, 23:506 (1984).

10. K. Gunther, J. Martens and M. Schickedanz, *Fresenius Z.Anal.Chem.*, 322:513 (1985).

11. U. A. Th. Brinkman and D. Kamminga, *J.Chromatogr.*, 330:375 (1985).

12. K. Gunther, J. Martens and M. Schickedanz, *Naturwissenschaften*, 72:149 (1985).

13. K. Gunther, J. Martens and M. Schickedanz, *Arch.Pharm.(Weinheim)*, 319, 461 (1986).

14. K. Gunther, J. Martens and M. Schickedanz, *Arch.Pharm.(Weinheim)*, 319:572 (1986).

15. K. Gunther, M. Schickedanz, K. Drauz and J. Martens, *Z.Anal.Chem.*, 325:298 (1986).

16. C. Syldatk, D. Cotoras, A. Moller and F. Wagner, *Biotech-Forum*, 3:9 (1986).

17. B. K. Vriesema, W.ten Hoeve, H. Wynberg and R. M. Kellogg, *Tetrahedron Lett.*, 26:2045 (1986).

18. R. Rausch *in* "Recent advances in thin-layer chromatography", eds. F.A.A. Dallas, H. Read, R.J. Ruane and I.D. Wilson, Plenum, 151 (1988).

19. S. Weinstein, *Tetrahedron Lett.*, 25:985 (1984).

20. N. Grinberg, *J.Chromatogr.*, 303:251 (1984).

21. F. Giess, "Fundamentals of thin-layer chromatography," Dr Alfred Heuthig Verlag, Heidlberg (1987).

APPENDIX 1 — CHIRAL CHROMATOGRAPHY LITERATURE 1987-1988

In order to compensate for the inevitable delay between the receipt of manuscripts and their publication in a final form a selection of highlights from the literature in the rapidly expanding field of chiral chromatography is provided (some 170 references) in the following pages. The material presented is mainly for the period 1987 to December 1988 and thus covers the year of the conference itself as well as providing details of those papers published during the production of the book.

Broadly speaking the chiral literature has been divided into three main subject areas, HPLC, GLC and TLC with additional minor categories such as SFC, Prep. and DCC considered in a small "miscellaneous" section. In the case of TLC, as the total literature on enantiomer separation is relatively small, all the references in our collection have been included, irrespective of their date of publication, to give what we believe to be a fairly complete coverage of this particular area.

The references are laid out in alphabetical order (by first author). In addition the references are indexed by author (all authors) and by subject to enable a rapid search for either a particular author or compound type etc.

Whilst, due to the rapidly expanding nature of the subject, this survey is undoubtedly incomplete and depends to no small extent on the idiosyncrasies of its compilers, we nevertheless hope that the information it contains will add to the usefulness of the volume as a whole.

R.J. Ruane and I.D. Wilson

HIGH-PERFORMANCE LIQUID CHROMATOGRAPHY/LIQUID CHROMATOGRAPHY (HPLC/LC)

1. Abidi S.L.
    OPTICAL RESOLUTION OF ROTENOIDS.
    J. Hetrocycl. Chem., 24, 845-52, (1987).

2. Alembik M.C., Wainer I.W.
    RESOLUTION AND ANALYSIS OF ENANTIOMERS OF AMPHETAMINES BY LIQUID
    CHROMATOGRAPHY ON A CHIRAL STATIONARY PHASE: COLLABORATIVE STUDY.
    J. Assoc. Off. Anal. Chem., 71, 530-3, (1988).

3. Armstrong D.W., Han Y.I., Han S.M.
    LIQUID CHROMATOGRAPHIC RESOLUTION OF ENANTIOMERS CONTAINIING
    SINGLE AROMATIC RINGS WITH β-CYCLODEXTRIN-BONDED PHASES.
    Anal. Chim. Acta, 208, 275-81, (1988).

4. Armstrong D.W., Spino L.A., Han S.M., Seeman J.I., Secor H.V.
    ENANTIOMERIC RESOLUTION OF RACEMIC NICOTINE AND NICOTINE ANALOGS BY
    MICROCOLUMN LIQUID CHROMATOGRAPHY WITH β-CYCLODEXTRIN INCLUSION
    COMPLEXES.
    J. Chromatogr., 411, 490-3, (1987).

5. Avgerinos A., Hutt A.J.
    DETERMINATION OF THE ENANTIOMERIC COMPOSITION OF IBUPROFEN IN
    HUMAN PLASMA BY HIGH PERFORMANCE LIQUID CHROMATOGRAPHY.
    J. Chromatogr., 415, 75-83, (1987).

6. Ballasteros P., Claramunt R.M., Elguero J., Gallego-Preciado M., Roussel C.,
    Chemlal A.
    ENANTIOMERIC RESOLUTION OF 3,5'-DIMETHYL-4,4'-DIBROMO-1,1'-
    BISPYRAZOLYLPHENYLMETHANE BY LIQUID CHROMATOGRAPHY ON
    TRIACETYLCELLULOSE.
    Heterocycles, 27, 351-6, (1988).

7. Bertucci C., Rosini C., Pini D., Salvadori P.
    CHIRAL STATIONARY PHASES AND CIRCULAR DICHROISM DETECTION IN HIGH
    PERFORMANCE LIQUID CHROMATOGRAPHY: DETERMINATION OF
    STEREOCHEMICAL PURITY OF DRUGS.
    J. Pharm. Biomed. Anal., 5, 171-6, (1987).

8. Blaschke G.
    SUBSTITUTED POLYACRYLAMIDES AS CHIRAL PHASES FOR THE RESOLUTION OF
    DRUGS.
    J. Chromatogr. Sci., 40, 179-98, (1988).

9. Blessington B., Crabb M., O'Sullivan J.
    CHIRAL HIGH PERFORMANCE LIQUID CHROMATOGRAPHIC STUDIES OF 2-(4-
    CHLORO-2-METHYLPHENOXY) PROPANOIC ACID.
    J. Chromatogr., 396, 177-82, (1987).

10. Bopp R.J., Kennedy J.H.
    PRACTICAL CONSIDERATIONS FOR CHIRAL SEPARATIONS OF PHARMACEUTICAL
    COMPOUNDS.
    LC-GC, 6, 514-22, (1988).

11. Bruegger R.R., Marti A.R., Meyer V.R., Arm H.
    OPTICAL RESOLUTION OF SAMPLES WITH WEAK INTERACTIONS ON CHIRAL
    "BRUSH TYPE" CHROMATOGRAPHIC STATIONARY PHASES.
    J. Chromatogr., 440, 197-207, (1988).

12. Caldwell J., Darbyshire J.F., Winter S.M., Hutt A.J.
    PITFALLS IN THE ENANTIOSELECTIVE ANALYSIS OF CHIRAL DRUGS. In
    Bioanalysis of Drugs and Metabolites, Especially Anti-Inflammatory and
    Cardiovascular. Edited by: Reid E., Robinson J.D., Wilson I.D.
    Plenum (1988), p 257-61.

13. Carunchio V., Messina A., Sinibalda M., Fanali S.
    HIGH-PERFORMANCE LIGAND-EXCHANGE CHROMATOGRAPHY OF AMINO ACIDS
    ON CHIRAL STATIONARY PHASES.
    J. High Resolut. Chromatogr. Chromatogr. Commun., 11, 401-4, (1988).

14. Coors C., Matusch R.
    SYNTHESIS OF CHIRAL STATIONARY PHASES FROM (S)-CAMPHOR-10-SULPHONYL
    DERIVATIVES FOR LIQUID CHROMATOGRAPHY OF DRUGS.
    J. High Resolut. Chromatogr. Chromatogr. Commun., 11, 422-3, (1988).

15. Corradini C., Federici F, Sinibaldi M., Messina A.
    HIGH PERFORMANCE LIGAND EXCHANGE CHROMATOGRAPHY ON CHIRAL
    DIAMINE-BONDED SILICA GEL.
    Chromatographia, 23, 118-20, (1987).

16. Dobashi A., Dobashi Y., Hara S.
    ENANTIOSELECTIVITY OF HYDROGEN-BOND ASSOCIATION IN LIQUID-SOLID
    CHROMATOGRAPHY.
    J. Liq. Chromatogr., 9, 243-67, (1986).

17. Dobashi A., Dobashi Y., Kinoshita K., Hara S.
    EXTENDED SCOPE OF ENANTIOMER RESOLUTION WITH CHIRAL DIAMIDE PHASES
    IN LIQUID CHROMATOGRAPHY.
    Anal. Chem., 60, 1985-7, (1988).

18. Dobashi Y., Hara S.
    A CHIRAL STATIONARY PHASE DERIVED FROM (R,R)-TARTRAMIDE WITH
    BROADENED SCOPE OF APPLICATION TO THE LIQUID CHROMATOGRAPHIC
    RESOLUTION OF ENANTIOMERS.
    J. Org. Chem., 52, 2490-6, (1987).

19. Edholm L.A., Lindberg C., Paulson J., Walhagen A.
DETERMINATION OF DRUG ENANTIOMERS IN BIOLOGICAL SAMPLES BY COUPLED
COLUMN LIQUID CHROMATOGRAPHY AND LIQUID CHROMATOGRAPHY-MASS
SPECTROMETRY.
J. Chromatogr., 424, 61-72, (1988).

20. Einarsson S., Folestad S., Josefsson B.
SEPARATION OF AMINO ACID ENANTIOMERS USING PRECOLUMN
DERIVATIZATION WITH O-PHTHALALDEHYDE AND 2,3,4,6-TETRA-O-ACETYL-1-
THIO-β-GLUCOPYRANOSIDE.
J. Liq. Chromatogr., 10, 1589-601, (1987).

21. Erlandsoon P., Hansson L., Isaksson R.
DIRECT ANALYTICAL AND PREPARATIVE RESOLUTION OF ENANTIOMERS USING
ALBUMIN ADSORBED TO SILICA AS A STATIONARY PHASE.
J. Chromatogr., 370, 475-83, (1986).

22. Feitsma K.G., Drenth B.F.H., Kooi K.H., Bosman J., De Zeeuw R.A.
ATTEMPTS TO OBTAIN SEPARATIONS OF CHIRAL ANTI CHOLINERGIC DRUGS
In Bioactive Analytes Including CNS Drugs, Peptides and Enantiomers.
Edited by Reid E., Scales B., Wilson I.D.
Plenum (1986), p 259-6.

23. Feitsma K.G., Drenth B.F.H., De Zeeuw R.A., Meijer D.K.F.
A NOTE ON CHIRAL DIFFERENCES IN THE DISPOSITION OF THE QUATERNARY
ANTICHOLINERGIC DRUG OXYPHENONIUM BROMIDE.
In Bioanalysis of Drugs and Metabolites, Especially Anti-Inflammatory and
Cardiovascular. Edited by: Reid E., Robinson J.D., Wilson I.D.
Plenum (1988), p 265-8.

24. Finn J.M.
RATIONAL DESIGN OF PIRKLE-TYPE CHIRAL STATIONARY PHASES.
J. Chromatogr. Sci., 40, 53-90, (1988).

25. Fujimoto Y., Ishi K., Nishi H., Tsumagari N., Kakimoto T., Shimizu R.
NEW DERIVATIZATION REAGENTS FOR THE RESOLUTION OF CARBOXYLIC ACID
ENANTIOMERS BY HIGH PERFORMANCE LIQUID CHROMATOGRAPHY.
J. Chromatogr., 402, 344-8, (1987).

26. Gargaro G., Gasparrini F., Misiti D., Palmieri G., Pierini M., Villani D.
NEW HPLC CHIRAL STATIONARY PHASES FOR ENANTIOMERIC RESOLUTION OF
SULFOXIDES AND SELENOXIDES.
Chromatographia, 24, 505-9, (1987).

27. Gazdag M., Szepesi G., Huszar L.
THE α-, β- AND γ-CYCLODEXTRINS AS MOBILE PHASE ADDITIVES IN THE HIGH
PERFORMANCE LIQUID CHROMATOGRAPHIC SEPARATION OF ENANTIOMERIC
COMPOUNDS.
J. Chromatogr., 436, 31-8, (1988).

28. Guebitz G., Juffmann F.
RESOLUTION OF THE ENANTIOMERS OF THYROID HORMONES BY HIGH
PERFORMANCE LIGAND-EXCHANGE CHROMATOGRAPHY USING A CHEMICALLY
BONDED CHIRAL STATIONARY PHASE.
J. Chromatogr., 404, 391-3, (1987).

29. Han S.M., Armstrong D.W.
USE OF MICROCOLUMN LIQUID CHROMATOGRAPHY WITH A CHIRAL STATIONARY
PHASE FOR THE SEPARATION OF LOW-RESOLUTION ENANTIOMERS.
J. Chromatogr., 389, 256-60, (1987).

30. Hayes S.M., Liu R.H., Tsang W.S., Legendre M.G., Berni R.J., Pillion D.J., Barnes S., Ho
M.H.
ENANTIOMERIC COMPOSITION ANALYSIS OF AMPHETAMINE AND
METHAMPHETAMINE BY CHIRAL PHASE HIGH PERFORMANCE LIQUID
CHROMATOGRAPHY-MASS SPECTROMETRY.
J. Chromatogr., 398, 239-46, (1987).

31. Hussenius A., Isaksson R., Matsson O.
SEPARATION OF ALKYL-SUBSTITUTED INDENE ENANTIOMERS BY
CHROMATOGRAPHY ON MICROCRYSTALLINE TRIACETYLCELLULOSE.
J. Chromatogr., 405, 155-62, (1987).

32. Hutt A.J., Caldwell J.
ENANTIOMERIC ANALYSIS OF 2-PHENYPROPIONIC ACID NSAID'S IN BIOLOGICAL
FLUIDS BY HPLC.
In Bioanalysis of Drugs and Metabolites, Especially Anti-Inflammatory and
Cardiovascular. Edited by: Reid E., Robinson J.D., Wilson I.D.
Plenum (1988), p 115-125.

33. Hyun M.H., Pirkle W.H.
PREPARATION AND EVALUATION OF A CHIRAL STATIONARY PHASE BEARING
BOTH π-ACIDIC AND -BASIC SITES.
J. Chromatogr., 393, 357-65, (1987).

34. Ichida A., Shibata T.
CELLULOSE DERIVATIVES AS STATIONARY CHIRAL PHASES.
J. Chromatogr. Sci., 40, 219-43, (1988).

35. Isaksson R., Sandstroem J., Eliaz M., Israely Z., Agranat I.
ENANTIOMER RESOLUTION OF NEFOPAM HYDROCHLORIDE, A NOVEL ANALGESIC:
A STUDY BY LIQUID CHROMATOGRAPHY AND CIRCULAR DICHROISM
SPECTROSCOPY.
J. Pharm. Pharmacol., 40, 48-50, (1988).

36. Iwaki K., Yoshida S., Nimura N., Kinoshita T., Takeda K., Ogura H.
ACTIVATED CARBAMATE REAGENT AS CHIRAL DERIVATIZING AGENT FOR LIQUID
CHROMATOGRAPHIC OPTICAL RESOLUTION OF ENANTIOMERIC AMINO
COMPOUNDS.
Chromatographia, 23, 899-902, (1986).

37. Iwaki K., Yoshida S., Nimura N., Kinoshita T., Takeda K., Ogura H.
PREPARATION OF CHIRAL STATIONARY PHASE VIA ACTIVATED CARBAMATE
INTERMEDIATE FOR LIQUID CHROMATOGRAPHIC OPTICAL RESOLUTION.
Chromatographia, 23, 727-30, (1987).

38. Johns D.
RESOLVING ISOMERS ON HPLC COLUMNS WITH CHIRAL STATIONARY PHASES.
Am. Lab., 19, 72-6, (1987).

39. Karnes H.T., Sarkar M.A.
ENANTIOMERIC RESOLUTION OF DRUG COMPOUNDS BY LIQUID
CHROMATOGRAPHY.
Pharm. Res., 4, 285-92, (1987).

40. Krueger G., Groetzinger J., Berndt H.
ENANTIOMERIC RESOLUTION OF AMINO ACID DERIVATIVES ON CHIRAL
STATIONARY PHASES BY HIGH PERFORMANCE LIQUID CHROMATOGRAPHY.
J. Chromatogr., 397, 223-32, (1987).

41. Ladanyi L., Sztruhar I., Slegel P., Vereczekey-Donath G.
DETERMINATION OF THE ENANTIOMERIC COMPOSITION OF CHIRAL CARBOXYLIC
ACIDS USING CHIRAL DERIVATIZATION AND HPLC.
Chromatographia, 24, 477-81, (1987).

42. Lalonde R.L., Bottorff M.B., Wainer I.W.
THE STUDY OF CHIRAL CARDIOVASCULAR DRUGS: ANALYTICAL APPROACHES
AND SOME PHARMACOLOGICAL CONSEQUENCES.
In Bioanalysis of Drugs and Metabolites, Especially Anti-Inflammatory and
Cardiovascular. Edited by: Reid E., Robinson J.D., Wilson I.D.
Plenum (1988), p 169-77.

43. Lam S., Malikin G.
RESOLUTION OF CHIRAL AMINES BY HIGH PERFORMANCE LIQUID
CHROMATOGRAPHY OF THEIR MIXED CHELATE COMPLEXES WITH COPPER(II)-L-
PROLINE.
J. Chromatogr., 368, 413-22, (1986).

44. Le Garrec L., Delee E., Pascal J.C., Jullien I.
DIRECT SEPARATION OF D- AND L-SOTALOL MANDELATE AND HYDROCHLORIDE
SALTS BY HIGH PERFORMANCE LIQUID CHROMATOGRAPHY.
J. Liq. Chromatogr., 10, 3015-23, (1987).

45. Lienne M., Caude M., Tambute A., Rosset R.
ENANTIOMERS SEPARATION BY LIQUID CHROMATOGRAPHY ON CHIRAL
STATIONARY PHASES.
Analusis, 15, 431-76, (1987).

46. Lipkowitz K.B., Demeter D.A., Parish C.A., Darden T.
ENANTIOSELECTIVE BINDING OF 2,2,2-TRIFLUORO-1-(9-ANTHYRL)ETHANOL ON A
CHIRAL STATIONARY PHASE: A THEORETICAL STUDY.
Anal. Chem., 59, 1731-3, (1987).

47. Lu X.L., Yang S.K.
RESOLUTION OF ENANTIOMERS AND DIASTEREOISOMERS BY HPLC IN
PHARMACEUTICAL ANALYSIS.
Yaoxue Xuebao, 23, 67-79, (1988).

48. McDaniel D.M., Snider B.G.
RESOLUTION OF α-ARYLACETIC ACID ENANTIOMERS ON TWO CHIRAL
STATIONARY PHASES.
J. Chromatogr., 404, 123-32, (1987).

49. McErlane K.M., Igwemezie L., Kerr C.R.
STEREOSELECTIVE ANALYSIS OF THE ENANTIOMERS OF MEXILETHINE BY HIGH
PERFORMANCE LIQUID CHROMATOGRAPHY USING FLUORESCENCE DETECTION
AND STUDY OF THEIR STEREOSELECTIVE DISPOSITION.
J. Chromatogr., 415, 335-46, (1987).

50. Mehta A.C.
DIRECT SEPARATION OF DRUG ENANTIOMERS BY HIGH PERFORMANCE LIQUID
CHROMATOGRAPHY WITH CHIRAL STATIONARY PHASES.
J. Chromatogr., 426, 1-13, (1988).

51. Miwa T., Ichikawa M., Tsuno M., Hattori T., Miyakawa T., Kayano M., Miyake Y.
DIRECT LIQUID CHROMATOGRAPHIC RESOLUTION OF RACEMIC COMPOUNDS. USE
OF OVOMUCOID AS A COLUMN LIGAND.
Chem. Pharm. Bull., 35, 682-6, (1987).

52. Mueller M.D., Bosshardt H.P.
ENANTIOMER RESOLUTION AND ASSAY OF PROPIONIC ACID-DERIVED
HERBICIDES IN FORMULATIONS BY USING CHIRAL LIQUID CHROMATOGRAPHY
AND ACHIRAL GAS CHROMATOGRAPHY.
J. Assoc. Off. Anal. Chem., 71, 614-17, (1988).

53. Nimura N., Iwaki K., Kinoshita T.
SEPARATION OF NOREPINEPHRINE ENANTIOMERS, DOPA ENANTIOMERS AND
DOPAMINE DERIVATIZED WITH O-PHTHALALDEHYDE-N-ACETYL-L-CYSTEINE BY
HIGH PERFORMANCE LIQUID CHROMATOGRAPHY.
J. Chromatogr., 402, 387-91, (1987).

54. Noctor T.A.G., Clark B.J., Fell A.F.
CHIRAL SEPARATION OF DRUG ENANTIOMERS BY HIGH PERFORMANCE LIQUID
CHROMATOGRAPHY.
Anal. Proc. (London), 23, 441-3, (1986).

55. Norinder U., Sundholm E.G.
THE USE OF COMPUTER AIDED CHEMISTRY TO PREDICT CHIRAL SEPARATION IN
LIQUID CHROMATOGRAPHY.
J. Liq. Chromatogr., 10, 2825-44, (1987).

56. Oi N., Kitahara H.
ENANTIOMER SEPARATION BY HPLC WITH SOME UREA DERIVATIVES OF L-
VALINE AS NOVEL CHIRAL STATIONARY PHASES.
J. Liq. Chromatogr., 9, 511-17, (1986).

57. Okamoto Y.
     SEPARATE OPTICAL ISOMERS BY CHIRAL HPLC.
     Chemtech, 17, 176-81, (1987).

58. Okamoto Y., Aburatani R., Fukumoto T., Hatada K.
     CHROMATOGRAPHIC RESOLUTION. XVII. USEFUL CHIRAL STATIONARY PHASES
     FOR HPLC. AMYLOSE TRIS(3,5-DIMETHYLPHENYLCARBAMATE) AND TRIS(3,5-
     DICHLOROPHENYLCARBAMATE) SUPPORTED ON SILICA GEL.
     Chem. Lett., 1857-60, (1987).

59. Okamoto Y., Aburatani R., Hatada K.
     CHROMATOGRAPHIC CHIRAL RESOLUTION. XIV. CELLULOSE TRIBENZOATE
     DERIVATIVES AS CHIRAL STATIONARY PHASES FOR HIGH PERFORMANCE LIQUID
     CHROMATOGRAPHY.
     J. Chromatogr., 389, 95-102, (1987).

60. Okamoto Y., Aburanti R., Kaida Y., Hatada K.
     DIRECT OPTICAL RESOLUTION OF CARBOXYLIC ACIDS BY CHIRAL HPLC ON
     TRIS(3,5-DIMETHYLPHENYLCARBAMATE)S OF CELLULOSE AND AMYLOSE.
     Chem. Soc. Jpn., Chem. Lett., 1125-8, (1988).

61. Okamoto Y., Hatada K.
     RESOLUTION OF ENANTIOMERS BY HPLC ON OPTICALLY ACTIVE
     POLY(TRIPHENYLMETHYL METHACRYLATE).
     J. Liq. Chromatogr., 9, 369-84, (1986).

62. Okamoto Y., Kawashima M., Hatada K.
     CHROMATOGRAPHIC RESOLUTION. XI. CONTROLLED CHIRAL RECOGNITION OF
     CELLULOSE TRIPHENYLCARBAMATE DERIVATIVES SUPPORTED ON SILICA GEL.
     J. Chromatogr., 363, 173-86, (1986).

63. Okamoto Y., Sakamoto H., Hatada K., Irie M.
     RESOLUTION OF ENANTIOMERS BY HPLC ON CELLULOSE TRANS- AND CIS-TRIS(4-
     PHENYLAZOPHENYLCARBAMATE).
     Chem. Lett., 983-6, (1986).

64. Okamoto Y., Hashima E., Ishikura M., Hatada K.
     THE CHIRAL RECOGNITION OF OPTICALLY ACTIVE POLY(TRIPHENYLMETHYL
     METHACRYLATE) DERIVATIVES AS STATIONARY PHASES FOR HPLC.
     Bull. Chem. Soc. Jpn., 61, 255-9, (1988).

65. Pedrazzini S., Zanoboni-Muciaccia W., Sacchi S., Forgione A.
     DETERMINATION OF FLUNOXAPROFEN ENANTIOMERS IN BIOLOGICAL FLUIDS BY
     HIGH PERFORMANCE LIQUID CHROMATOGRAPHY.
     J. Chromatogr., 415, 214-20, (1987).

66. Pettersson C.
     LIQUID CHROMATOGRAPHIC SEPARATION OF ENANTIOMERS USING CHIRAL
     ADDITIVES IN THE MOBILE PHASE.
     Trends in Anal. Chem., 7, 209-17, (1988).

67. Pettersson C., Arvidsson T., Karlsson A.L., Marle I.
CHROMATOGRAPHIC RESOLUTION OF ENANTIOMERS USING ALBUMIN AS
COMPLEXING AGENT IN THE MOBILE PHASE.
J. Pharm. Biomed. Anal., 4, 221-35, (1986).

68. Pettersson C., Gioeli C.
SEPARATION OF ENANTIOMERIC ACIDS USING IMMOBOLIZED ACETYLQUININE
AS A CHIRAL STATIONARY PHASE.
J. Chromatogr., 398, 247-54, (1987).

69. Pettersson C., Schill G.
ENANTIOMER SEPARATION IN ION-PAIRING SYSTEMS.
J. Chromatogr. Sci., 40, 283-313, (1988).

70. Pflugman G., Spahn H., Mutschler E.
DETERMINATION OF METOPROLOL ENANTIOMERS IN PLASMA AND URINE USING
(S)-(-)-PHENYLETHYL ISOCYANATE AS A CHIRAL REAGENT.
J. Chromatogr., 421, 161-4. (1987).

71. Pini D., Rosini C., Bertucci C., Altemura P., Salvadori P.
CHROMATOGRAPHIC RESOLUTION OF ALKYL ARYL CARBINOLS ON A SILICA-
SUPPORTED QUININE CHIRAL STATIONARY PHASE.
Gazz. Chim. Ital., 116, 603-6, (1986).

72. Pirkle W.H., Alessi D.M., Hyun M.H., Pochapsky T.C.
SEPARATION OF SOME ENANTIOMERIC DI- AND TRIPEPTIDES ON CHIRAL
STATIONARY PHASES.
J. Chromatogr., 398, 203-9, (1987).

73. Pirkle W.H., Pochapsky T.C.
CHIRAL STATIONARY PHASES FOR THE DIRECT LC SEPARATION OF
ENANTIOMERS.
Adv. Chromatogr., 27, 73-127, (1987).

74. Pirkle W.H., Sowin T.J.
SYNTHESIS AND SEPARATION OF DIASTEREOMERIC IMINO ALCOHOL
DERIVATIVES OF CHIRAL PHTHALIDES: A METHOD FOR ASSIGNMENT OF
PHTHALIDE ABSOLUTE CONFIGURATIONS.
J. Org. Chem., 52, 3011-17, (1987).

75. Prinsen W.J.C., Laarhoven W.H.
DETERMINATION OF THE "ENANTIOMERIC EXCESS" OF HEXAHELICENE AND ITS
METHYL-SUBSTITUTED DERIVATIVES BY HIGH PERFORMANCE LIQUID
CHROMATOGRAPHY.
J. Chromatogr., 393, 377-90, (1987).

76. Railton I.D.
RESOLUTION OF THE ENANTIOMERS OF ABSCISIC ACID METHYL ESTER BY HIGH
PERFORMANCE LIQUID CHROMATOGRAPHY USING A STATIONARY PHASE OF
CELLULOSE TRIS(3,5-DIMETHYLPHENYLCARBAMATE)-COATED SILICA GEL.
J. Chromatogr., 402, 371-3, (1987).

77. Ravichandran K., Rogers L.B.
CHIRAL SEPARATIONS OF AMINO ACIDS USING DI- AND TRI-PEPTIDES FOR
ZWITTER-ION PAIR CHROMATOGRAPHY.
J. Chromatogr., 402, 49-54, (1987).

78. Seeman J.I., Secor H.V., Armstrong D.W., Timmons K.D., Ward T.J.
ENANTIOMERIC RESOLUTION AND CHIRAL RECOGNITION OF RACEMIC NICOTINE
AND NICTOINE ANALOGS BY β-CYCLODEXTRIN COMPLEXATION STRUCTURE-
ENANTIOMERIC RESOLUTION. RELATIONSHIPS IN HOST-GUEST INTERACTIONS.
Anal. Chem., 60, 2120-7, (1988).

79. Shimizu T., Kobayashi M.
OPTICAL RESOLUTION OF ASYMEMTRIC SELENOXIDES BY HIGH PERFORMANCE
LIQUID CHROMATOGRAPHY USING OPTICALLY ACTIVE COLUMN.
Bull. Chem. Soc. Jpn., 59, 2654-6, (1986).

80. Shinbo T., Yamaguchi T., Nishimura K., Sugiura M.
CHROMATOGRAPHIC SEPARATION OF RACEMIC AMINO ACIDS BY USE OF CHIRAL
CROWN ETHER-COATED REVERSED-PHASE PACKINGS.
J. Chromatogr., 405, 145-53, (1987).

81. Sinibaldi M., Federici F., Fanali S., Messina A.
HIGH PERFORMANCE LIQUID CHROMATOGRAPHY RESOLUTION OF RACEMATES
USING A CHIRAL ADDITIVE TO THE ELUENT.
J. High Resolut. Chromatogr. Chromatogr. Commun., 10, 206-7, (1987).

82. Sinibaldi M., Federici F., Messina A.
LIQUID CHROMATOGRAPHIC RESOLUTION OF ENANTIOMERS ON CHIRAL AMINE
DERIVATIVE-BONDED SILICA GEL.
Ann. Chim. (Rome), 77, 889-98, (1987).

83. Straka R.J., Lalonde R.L., Wainer I.W.
MEASUREMENT OF UNDERIVATIZED PROPRANOLOL ENANTIOMERS IN SERUM
USING A CELLULOSE-TRIS(3,5-DIMETHYLPHENYLCARBAMATE) HIGH
PERFORMANCE LIQUID CHROMATOGRAPHY (HPLC) CHIRAL STATIONARY PHASE.
Pharm. Res., 5, 187-9, (1988).

84. Strasak M., Bystricky S.
PARTIAL RESOLUTION OF COBALT (III) CHELATE COMPLEXES BY GEL PERMEATION
CHROMATOGRAPHY.
J. Chromatogr., 403, 331-5, (1987).

85. Sybilska D., Zukowski J., Bojarski J.
RESOLUTION OF MEPHENYTOIN AND SOME CHIRAL BARBITURATES INTO
ENANTIOMERS BY REVERSED-PHASE HIGH PERFORMANCE LIQUID
CHROMATOGRAPHY VIA β-CYCLODEXTRIN INCLUSION COMPLEXES.
J. Liq. Chromatogr., 9, 591-606, (1986).

86. Takagi T., Itabashi Y.
RAPID SEPARATIONS OF DIACYL- AND DIALKYLGLYCEROL ENANTIOMERS BY
HIGH PERFORMANCE LIQUID CHROMATOGRAPHY ON A CHIRAL STATIONARY
PHASE.
Lipids, 22, 596-600, (1987).

87. Tambute A., Bergos A., Lienne M., Caude M., Rosset R.
     NEW CHIRAL STATIONARY PHASES CONTAINING A PHOSPHOROUS ATOM AS
     ASYMMETRIC CENTER. I. SYNTHESIS AND FIRST CHROMATOGRAPHIC RESULTS.
     J. Chromatogr., 396, 65-81, (1987).

88. Vloon W.J., Siekerman C., Kraak J.C.
     L-N-(3,5-DIMETHOXYBENZOYL)ISOLEUCINE CHIRAL STATIONARY PHASE FOR
     SEPARATION OF ENANTIOMERS BY HPLC.
     Chromatographia, 24, 655-8, (1987).

89. Wagner J., Gaget C., Heintzelmann B., Wolf E.
     RESOLUTION OF THE ENANTIOMERS OF VARIOUS α-SUBSTITUTED ORNITHINE
     AND LYSINE ANALOGS BY HIGH PERFORMANCE LIQUID CHROMATOGRAPHY WITH
     CHIRAL ELUENT AND BY GAS CHROMATOGRAPHY ON CHIRASIL-VAL.
     Anal. Biochem., 164, 102-16, (1987).

90. Wagner J., Wolf E., Heintzelmann B., Gaget C.
     CHIRAL SEPARATION OF ENANTIOMERS OF SUBSTITUTED α- AND β-ALANINE
     AND γ-AMINOBUTYRIC ACID ANALOGS BY GAS CHROMATOGRAPHY AND HIGH
     PERFORMANCE LIQUID CHROMATOGRAPHY.
     J. Chromatogr., 392, 211-24, (1987).

91. Wainer I.W.
     A PRACTICAL GUIDE TO THE SELECTION AND USE OF HPLC CHIRAL STATIONARY
     PHASES.
     Published by: Baker J.T., (1988).

92. Wainer I.W.
     PROPOSALS FOR THE CLASSIFICATION OF HIGH PERFORMANCE LIQUID
     CHROMATOGRAPHIC CHIRAL STATIONARY PHASES: HOW TO CHOOSE THE RIGHT
     COLUMN.
     Trends Anal. Chem., 6, 125-34, (1987).

93. Wainer I.W., Alembik M.C.
     THE ENANTIOMERIC RESOLUTION OF BIOLOGICALLY ACTIVE MOLECULES ON
     COMMERCIALLY AVAILABLE LIQUID CHROMATOGRAPHIC CHIRAL STATIONARY
     PHASES.
     J. Chromatogr. Sci., 40, 355-84, (1988).

94. Wainer I.W., Alembik M.C., Smith E.
     RESOLUTION OF ENANTIOMERIC AMIDES ON A CELLULOSE TRIBENZOATE CHIRAL
     STATIONARY PHASE. MOBILE PHASE MODIFIER EFFECTS ON RETENTION AND
     STEREOSELECTIVITY.
     J. Chromatogr., 388, 65-74, (1987).

95. Wainer I.W., Stiffin R.M., Shibata T.
     RESOLUTION OF ENANTIOMERIC AROMATIC ALCOHOLS ON A CELLULOSE
     TRIBENZOATE HIGH PERFORMANCE LIQUID CHROMATOGRAPHY CHIRAL
     STATIONARY PHASE. A PROPOSED CHIRAL RECOGNITION MECHANISM.
     J. Chromatogr., 411, 139-51, (1987).

96.   Weaner L.E., Hoerr D.C.
      SEPARATION OF FATTY ACID ESTER AND AMIDE ENANTIOMERS BY HIGH
      PERFORMANCE LIQUID CHROMATOGRAPHY ON CHIRAL STATIONARY PHASES.
      J. Chromatogr., 437, 109-19, (1988).

97.   Whelpton R., Jonas G., Buckley D.G.
      HIGH PERFORMANCE LIQUID CHROMATOGRAPHIC RESOLUTION OF THE
      ENANTIOMERS OF THIORIDAZINE, ITS METABOLITES AND RELATED
      COMPOUNDS.
      J. Chromatogr., 426, 223-8, (1988).

98.   Wuis E.W., Beneken-Kolmer E.W.J., Van Beijsterveldt L.E.C., Van der Kleyn E.
      ENANTIOSELECTIVE HIGH PERFORMANCE LIQUID CHROMATOGRAPHIC
      DETERMINATION OF BACLOFEN AFTER DERIVATIZATION WITH A CHIRAL
      ADDUCT OF O-PHTHALDIALDEHYDE.
      J. Chromatogr., 415, 419-22, (1987).

99.   Wulff G., Minarik M.
      ENZYME-ANALOGUE BUILT POLYMERS PART XX. PRONOUNCED EFFECT OF
      TEMPERATURE ON RACEMIC RESOLUTION USING TEMPLATE-IMPRINTED
      POLYMERIC SORBENTS.
      J. High Resolut. Chromatogr. Chromatogr. Commun., 9, 607-8, (1986).

100. Wulff G., Minarik M.
      TAILOR-MADE SORBENTS. A MODULAR APPROACH TO CHIRAL SEPARATIONS.
      J. Chromatogr. Sci., 40, 15-52, (1988).

101. Yamashita J., Numakura T., Kita H., Suzuki T., Oi S., Miyano S., Hashimoto H.,
      Takai N.
      HIGH PERFORMANCE LIQUID CHROMATOGRAPHIC SEPARATION OF
      ENANTIOMERS ON AXIALLY CHIRAL BINAPHTHALENE DERIVATIVES BONDED TO
      SILICA GEL.
      J. Chromatogr., 403, 275-9, (1987).

102. Yang Q., Sun Z., Ling D.
      RESOLUTION OF ENANTIOMERIC DRUGS OF SOME β-AMINO ALCOHOLS AS THEIR
      UREA DERIVATIVES BY HIGH PERFORMANCE LIQUID CHROMATOGRAPHY.
      J. Chromatogr., 447, 208-11, (1988).

103. Yang. S.K., Mushtaq M., Fu P.P.
      ELUTION ORDER-ABSOLUTE CONFIGURATION RELATIONSHIP OF K-REGION
      DIHYDRODIOL ENANTIOMERS OF BENZ[A]ANTHRACENE DERIVATIVES IN CHIRAL
      STATIONARY PHASE HIGH PERFORMANCE LIQUID CHROMATOGRAPHY.
      J. Chromatogr., 371, 195-209, (1986).

104. Yang S.K., Mushtaq M., Weems H.B., Fu P.P.
      CHIRAL RECOGNITION MECHANISMS IN THE DIRECT RESOLUTION OF DIOL
      ENANTIOMERS OF SOME POLYCYCLIC AROMATIC HYDROCARBONS BY HIGH
      PERFORMANCE LIQUID CHROMATOGRAPHY WITH CHIRAL STATIONARY PHASES.
      J. Liq. Chromatogr., 9, 473-92, (1986).

105. Yuki Y., Saigo K.,Kimoto H., Tachibana K., Hasegawa M.
NOVEL CHIRAL STATIONARY PHASES FOR OPTICAL RESOLUTION BY LIGAND-
EXCHANGE HIGH PERFORMANCE LIQUID CHROMATOGRAPHY.
J. Chromatogr., 400, 65-75, (1987).

106. Yuki Y., Saigo K., Tachibana K., Hasegawa M.
NOVEL CHIRAL STATIONARY PHASES FOR THE RESOLUTION OF THE
ENANTIOMERS OF AMINO ACIDS BY LIGAND EXCHANGE CHROMATOGRAPHY.
Chem. Lett., 1347-50, (1986).

107. Zief M.
INFLUENCE OF THE MOBILE PHASE ON CHIRAL LIQUID CHROMATOGRAPHY
SEPARATIONS.
J. Chromatogr. Sci., 40, 315-35, (1988).

108. Zukowski J., Sybilska D., Bojarski J.
APPLICATION OF α- AND β-CYCLODEXTRIN AND HEPTAKIS (2,6-DI-O-METHYL)-β-
CYCLODEXTRIN AS MOBILE PHASE COMPONENTS FOR THE SEPARATION OF SOME
CHIRAL BARBITURATES INTO ENANTIOMERS BY REVERSED-PHASE HIGH
PERFORMANCE LIQUID CHROMATOGRAPHY.
J. Chromatogr., 364, 225-32, (1986).

109. Zukowski J., Sybilska D.,Bojarski J., Szejtli J.
RESOLUTION OF CHIRAL BARBITURATES INTO ENANTIOMERS BY REVERSED-
PHASE HIGH PERFORMANCE LIQUID CHROMATOGRAPHY USING METHYLATED β-
CYCLODEXTRINS.
J. Chromatogr., 436, 381-90, (1988).

GAS LIQUID CHROMATOGRAPHY (GC)

1. Alexander G., Juvancz Z., Szejtli J.
CYCLODEXTRINS AND THEIR DERIVATIVES AS STATIONARY PHASES IN GC
CAPILLARY COLUMNS
J. High Resolut. Chromatogr. Chromatogr. Commun., 11, 110-13, (1988).

2. Bradshaw J.S., Aggarwal S.K., Rouse C.A., Tarbet B.J., Markides K.E., Lee M.L.
POLYSILOXANES CONTAINING THERMALLY STABLE CHIRAL AMIDE SIDE-CHAINS
FOR CAPILLARY GAS AND SUPERCRITICAL FLUID CHROMATOGRAPHY.
J. Chromatogr., 405, 169-77, (1987).

3. Beck O., Repke D.B., Faull K.F.
6-HYDROXYMETHTRYPTOLINE IS NATURALLY OCCURRING IN MAMMALIAN
URINE: IDENTIFICATION BY COMBINED CHIRAL CAPILLARY GAS
CHROMATOGRAPHY AND HIGH RESOLUTION MASS SPECTROMETRY.
Biomed. Environ. Mass Spectrom., 13, 469-72, (1986).

4. Coutts R.T., Pasutto F.M., Jamali F., Singh N.
A FACILE GAS CHROMATOGRAPHIC METHOD FOR RESOLVING THE ENANTIOMERS
OF SEVERAL NON STEROIDAL ANTI-INFLAMMATORY DRUGS.
Dev. Drugs Mod. Med., 232-6, (1986).

5.   Davies B.E.
      DEVELOPMENT OF A CHIRAL CAPILLARY GC METHOD FOR THE QUANTITATION OF
      THE ENANTIOMERS OF CROMAKALIM IN BIOLOGICAL FLUIDS.
      In Bioanalysis of Drugs and Metabolites, Especially Anti-Inflammatory and
      Cardiovascular. Edited by: Reid E., Robinson J.D., Wilson I.D.
      Plenum (1988), p 179-83.

6.   Degenhardt C.E.A.M., Van den Berg G.R., De Jong L.P.A., Benschop H.P., Van Genderen
      J., Van de Meent D.
      ENANTIOSPECIFIC COMPLEXATION GAS CHROMATOGRAPHY OF NERVE AGENTS.
      ISOLATION AND PROPERTIES OF THE ENANTIOMERS OF ETHYL N,N-
      DIMETHYLPHOSPHORAMIDOCYANIDATE (TABUN).
      J. Am. Chem. Soc., 108, 8290-1, (1986).

7.   Ernst-Cabrera K., Koenig W.A.
      ENANTIOMERIC ANALYSIS BY GAS CHROMATOGRAPHY ON CHIRAL
      POLYSILOXANES.
      React. Polym. Ion Exch. Sorbents, 6, 267-74, (1988).

8.   Frank H.
      SEPARATION OF CHIRAL DRUGS AND METABOLITES BY CAPILLARY GAS
      CHROMATOGRAPHY.
      In Bioactive Analytes Including CNS Drugs, Peptides and Enantiomers. Edited by: Reid
      E.R., Scales B., Wilson I.D.
      Plenum (1986), p 285-7.

9.   Gaget C., Wolf E., Heintzelmann B., Wagner J.
      SEPARATION OF THE ENANTIOMERS OF SUBSTITUTED PUTRESCINE AND
      CADAVERINE ANALOGS BY GAS CHROMATOGRAPHY ON CHIRAL AND ACHIRAL
      STATIONARY PHASES.
      J. Chromatogr., 395, 597-608, (1987).

10.   Gray A., Mathews B.L., Self C.
      ANALYSIS OF THE HERBICIDE FLAMPROP M-ISOPROPYL (SINGLE ENANTIOMER)
      USING CAPILLARY GAS CHROMATOGRAPHY.
      Pestic. Sci. Biotechnol., Proc. 6th Int. Congr. Pestic. Chem., 249-52. Edited by:
      Greenhalgh R., Roberts T.R.
      Blackwell, Oxford, UK, (1987).

11.   Gyllenhaal O., Lamm B., Vessman J.
      ENANTIOMER SEPARATION OF TOCAINIDE AND SOME OF ITS ANALOGS AND
      DERIVATIVES ON A CHIRASIL-VAL CAPILLARY COLUMN.
      J. Chromatogr., 411, 285-95, (1987).

12.   Halterman R.L., Roush W.R., Hoong L.K.
      ANALYTICAL RESOLUTION OF SECONDARY METHYL ETHERS BY CHIRAL
      COMPLEXATION GAS CHROMATOGRAPHY.
      J. Org. Chem., 52, 1152-5, (1987).

13.   Hara S., Okabe H., Mihashi K.
      GAS-LIQUID CHROMATOGRAPHIC SEPARATION OF ALDOSE ENANTIOMERS AS
      TRIMETHYLSILYL ETHERS OF METHYL 2-(POLYHYDROXYALKYL)THIAZOLIDINE-
      4(R)-CARBOXYLATES.
      Chem. Pharm. Bull., 35, 501-6, (1987).

14. Jemal M., Cohen A.I.
ENANTIOMERIC PURITY DETERMINATION OF L-PROLINE BENZYL ESTER BY
CHIRAL COLUMN GAS CHROMATOGRAPHY.
J. Chromatogr., 392, 442-6, (1987).

15. Koenig W.A.
THE PRACTICE OF ENANTIOMER SEPARATION BY CAPILLARY GAS
CHROMATOGRAPHY.
Dr. Alfred Huethig Verlag: Heidelberg, 104 pp., (1987).

16. Koppenhoefer B., Allmendinger H.
DERIVATIZATION REACTIONS FOR ENANTIOMER RESOLUTION OF HYDROXY
ACIDS BY GAS CHROMATOGRAPHY: A COMPARISON OF METHODS.
Fresenius' Z. Anal. Chem., 326, 434-40, (1987).

17. Koppenhoefer B., Allmendinger H.
DIRECT ENANTIOMER RESOLUTION OF PRIMARY, SECONDARY AND TERTIARY
ALCOHOLS BY GAS CHROMATOGRAPHY ON CHIRASIL-VAL.
Chromatographia, 21, 503-8, (1986).

18. Koppenhoefer B., Allmendinger H., Bayer E.
ENANTIOMER AND DIASTEREOMER RESOLUTION OF DIPEPTIDES BY GAS
CHROMATOGRAPHY ON CHIRASIL-VAL.
J. High Resolut. Chromatogr. Chromatogr. Commun., 10, 324-6, (1987).

19. Koppenhoefer B., Koch E.M., Nicholson G.J., Bayer E.
CHIRAL RECOGNITION OF HALO CARBOXYLIC ACID AMIDES STUDIED BY GAS
CHROMATOGRAPHY ON CHIRASIL-VAL.
J. Chromatogr., 406, 157-66, (1987).

20. Koscielski T., Sybilska D., Jurczak J.
SEPARATION PROCESSES IN GAS-LIQUID CHROMATOGRAPHY BASED ON
FORMATION OF α-CYCLODEXTRIN-CHIRAL HYDROCARBONS INCLUSION
COMPLEXES.
J. Inclusion Phenom., 5, 69-72, (1987).

21. Rieck M., Hagen M., Lutz S., Koenig W.A.
ENANTIOMER SEPARATION OF ACYLOINS AND CYANOHYDRINS BY
ENANTIOSELECTIVE CAPILLARY GAS CHROMATOGRAPHY.
J. Chromatogr., 439, 301-6, (1988).

22. Schurig V.
ENANTIOMER SEPARATION BY COMPLEXATION GAS CHROMATOGRAPHY -
APPLICATIONS IN CHIRAL ANALYSIS OF PHEROMONES AND FLAVORS.
In Bioflavour '87, Proc. Int. Conf., 35-54. Edited by: Schreier P., de Gruyter, Berlin, Fed.
Rep. Ger. (1988).

23. Schurig V.
SEMI-PREPARATIVE ENANTIOMER SEPARATION OF 1,6-
DIOXASPIRO[4,4]NONANES BY COMPLEXATION GAS CHROMATOGRAPHY.
Naturwissenschaften, 74, 190-1, (1987).

24. Schurig V., Ossig A., Link R.
     ENANTIOMER SEPARATION OF 2-HALOCARBOXYLIC ACID ESTERS BY CHIRAL
     COMPLEXATION GAS CHROMATOGRAPHY.
     J. High Resolut. Chromatogr. Chromatogr. Commun., 11, 89-93, (1988).

25. Singh N.N., Pasutto F.M., Coutts R.T., Jamali F.
     GAS CHROMATOGRAPHIC SEPARATION OF OPTICALLY ACTIVE ANTI-
     INFLAMMATORY 2-ARYLPROPIONIC ACIDS USING (+)- OR (-)-AMPHETAMINE AS
     DERIVATIZING REAGENT.
     J. Chromatogr., 378, 125-35, (1986).

26. Wagner J., Gaget C., Heintzelmann B., Wolf E.
     RESOLUTION OF THE ENANTIOMERS OF VARIOUS α-SUBSTITUTED ORNITHINE
     AND LYSINE ANALOGS BY HIGH PERFORMANCE LIQUID CHROMATOGRAPHY WITH
     CHIRAL ELUANT AND BY GAS CHROMATOGRAPHY.
     Anal. Biochem., 164, 102-16, (1987).

27. Wagner J., Wolf E., Heintzelmann B., Gaget C.
     CHIRAL SEPARATION OF ENANTIOMERS OF SUBSTITUTED α- AND β-ALANINE
     AND γ-AMINOBUTYRIC ACID ANALOGS BY GAS CHROMATOGRAPHY AND HIGH
     PERFORMANCE LIQUID CHROMATOGRAPHY.
     J. Chromatogr., 392, 211-24, (1987).

THIN-LAYER CHROMATOGRAPHY (TLC)

1.   Alak A., Armstrong D.W.
     THIN-LAYER CHROMATOGRAPHIC SEPARATION OF OPTICAL, GEOMETRICAL AND
     STRUCTURAL ISOMERS.
     Anal. Chem., 58, 582-4, (1986).

2.   Armstrong D.W., Faulkner J.R., Han S.M.
     USE OF HYDROXYPROPYL- AND HYDROXYETHYL-DERIVATIZED β-CYCLODEXTRINS
     FOR THE THIN LAYER CHROMATOGRAPHIC SEPARATION OF ENANTIOMERS AND
     DIASTEREOMERS.
     J. Chromatogr., 452, 323-30, (1988).

3.   Bhushan R., Ali I.
     RESOLUTION OF ENANTIOMERIC AMINO ACIDS ON BERBERINE-IMPREGNATED
     SILICA PLATES.
     Fresenius' Z. Anal. Chem., 329, 793, (1988).

4.   Bhushan R., Ali I.
     RESOLUTION OF ENANTIOMERIC MIXTURES OF PHENYLTHIOHYDANTOIN AMINO
     ACIDS ON (+)-TARTARIC ACID-IMPREGNATED SILICA GEL PLATES.
     J. Chromatogr., 392, 460-3, (1987).

5.   Brinkman U.A.Th., Kamminga D.
     RAPID SEPARATION OF ENANTIOMERS BY THIN-LAYER CHROMATOGRAPHY ON A
     CHIRAL STATIONARY PHASE.
     J. Chromatogr., 410, 226-9, (1987).

6. Feldberg R.S., Repucci L.M.
    RAPID SEPARATION OF ANOMERIC PURINE NUCLEOSIDES BY THIN-LAYER
    CHROMATOGRAPHY ON A CHIRAL STATIONARY PHASE.
    J. Chromatogr., 410, 226-9, (1987).

7. Gont L.K., Neuendorf S.K.
    ENANTIOMERIC SEPARATION OF N-CARBAMYLTRYPTOPHAN BY THIN-LAYER
    CHROMATOGRAPHY ON A CHIRAL STATIONARY PHASE.
    J. Chromatogr., 391, 343-5, (1987).

8. Grinberg N., Weinstein S.
    ENANTIOMERIC SEPARATION OF DNS-AMINO ACIDS BY REVERSED-PHASE THIN-
    LAYER CHROMATOGRAPHY.
    J. Chromatogr., 303, 251-5, (1984).

9. Gunther K., Martens J., Schickedanz M.
    THIN-LAYER CHROMATOGRAPHIC ENANTIOMERIC RESOLUTION VIA LIGAND
    EXCHANGE.
    Angew. Chem. Int. Ed. Engl., 23, 506, (1984).

10. Gunther K., Schickedanz M., Drauz K., Martens J.
    THIN-LAYER CHROMATOGRAPHIC ENANTIOMERIC RESOLUTION OF α-ALKYL
    AMINO ACIDS.
    Fresenius' Z. Anal. Chem., 325, 298-9, (1986).

11. Gunther K., Schickedanz M., Martens J.
    THIN-LAYER CHROMATOGRAPHIC ENANTIOMERIC RESOLUTION.
    Naturwissenschaften, 72, 149-50, (1985).

12. Gunther K., Rausch R.
    THIN-LAYER CHROMATOGRAPHIC ENANTIOMERIC RESOLUTION BASED ON
    LIGAND EXCHANGE CHROMATOGRAPHY.
    In Proc. Int. Symp. Instrum. High Perform. Thin-Layer Chromatogr., 3rd, 469-74.
    Edited by: Kaiser R.E., Inst. Chromatogr., Bad Durkheim, Fed. Rep. Ger., (1985).

13. Rausch R.
    THE IMPORTANCE OF THIN-LAYER CHROMATOGRAPHY AS A RAPID METHOD FOR
    THE CONTROL OF OPTICAL PURITY.
    In Recent Advances in Thin-layer Chromatography. Edited by: Dallas F.A.A., Read H.,
    Ruane R.J., Wilson I.D.
    Plenum (1988), p 151-61.

14. Slegel P., Vereczkey-Donath G., Ladanyi L., Toth-Lauritz M.
    ENANTIOMERIC SEPARATION OF CHIRAL CARBOXYLIC ACIDS, AS THEIR
    DIASTEREOMERIC CARBOXAMIDES, BY THIN-LAYER CHROMATOGRAPHY.
    J. Pharm. Biomed. Anal., 5, 665-73, (1987).

15. Wainer I.W., Bruner C.A., Doyle T.D.
    DIRECT RESOLUTION OF ENANTIOMERS VIA THIN-LAYER CHROMATOGRAPHY
    USING A CHIRAL ADSORBENT.
    J. Chromatogr., 264, 154, (1983).

16. Weinstein S.
RESOLUTION OF OPTICAL ISOMERS BY THIN-LAYER CHROMATOGRAPHY.
Tetrahedron Lett., 25, 985-6, (1984).

17. Wilson I.D.
TOWARDS CHIRAL TLC PLATES: SOME PRELIMINARY STUDIES.
Bioactive Analytes Including CNS Drugs, Peptides and Enantiomers. Edited by: Reid E.,
Scales B., Wilson I.D.
Plenum (1986), p 277-81.

18. Yuasa S., Shimada A., Kameyama K., Yasui M., Adzuma K.
CELLULOSE THIN-LAYER AND COLUMN CHROMATOGRAPHY FOR RESOLUTION OF
DL-TRYPTOPHAN.
J. Chromatogr. Sci., 18, 311-4, (1980).

GENERAL/REVIEWS (GEN REV)

1. Allenmark S.G.
CHROMATOGRAPHIC ENANTIOSEPARATION: METHODS AND APPLICATIONS.
Ellis Horwood, (1988).

2. Bopp R.J., Kennedy J.H.
PRACTICAL CONSIDERATIONS FOR CHIRAL SEPARATIONS OF PHARMACEUTICAL
COMPOUNDS.
LC-GC, 6, 514-22, (1988).

3. Kinkel J.N., Fraenkel W., Blaschke G.
THE SEPARATION OF ENANTIOMERS ON POLYMER COATED CHIRAL SILICA GELS.
Kontakte, 3-14, (1987).

4. Koenig W.A.
THE PRACTICE OF ENANTIOMER SEPARATION BY CAPILLARY GAS
CHROMATOGRAPHY.
Dr. Alfred Huethig Verlag: Heidelberg, (1987).

5. BIOACTIVE ANALYTES INCLUDING CNS DRUGS, PEPTIDES AND ENANTIOMERS.
Edited by Reid E., Scales B., Wilson I.D. Plenum (1986).

6. Wainer I.W.
A PRACTICAL GUIDE TO THE SELECTION AND USE OF HPLC CHIRAL STATIONARY
PHASES.
Baker J.T., (1988).

7. Wainer I.W.
PROPOSALS FOR THE CLASSIFICATION OF HIGH PERFORMANCE LIQUID
CHROMATOGRAPHIC CHIRAL STATIONARY PHASES: HOW TO CHOOSE THE RIGHT
COLUMN.
Trends Anal. Chem., 6, 125-34, (1987).

DROPLET COUNTER-CURRENT CHROMATOGRAPHY (DCC)

1. Oya S., Snyder J.K.
    CHIRAL RESOLUTION OF A CARBOXYLIC ACID USING DROPLET COUNTER-
    CURRENT CHROMATOGRAPHY.
    J. Chromatogr., 370, 333-8, (1986).

PREPARATIVE CHROMATOGRAPHY (PREP)

1. Erlandsoon P., Hansson L., Issaksson R.
    DIRECT ANALYTICAL AND PREPARATIVE RESOLUTION OF ENANTIOMERS USING
    ALBUMIN ADSORBED TO SILICA AS A STATIONARY PHASE.
    J. Chromatogr., 370, 475-83, (1986).

2. Meyer V.R.
    SOME ASPECTS OF THE PREPARATIVE SEPARATION OF ENANTIOMERS ON
    CHIRAL STATIONARY PHASES.
    Chromatographia, 24, 639-45, (1987).

3. Perry J.A., Rateike J.D., Szczerba T.J.
    PREPARATIVE CHROMATOGRAPHY AND PURIFICATION.
    J. Liq. Chromatogr., 9, 3297-309, (1986).

4. Schurig V.
    SEMI-PREPARATIVE ENANTIOMER SEPARATION OF 1,6-
    DIOXASPIRO[4,4]NONANES BY COMPLEXATION GAS CHROMATOGRAPHY.
    Naturwissenschaften, 74, 190-1, (1987).

SUPERCRITICAL FLUID CHROMATOGRAPHY (SFC)

1. Bradshaw J.S., Aggerwal S.K., Rouse C.A., Tarbet B.J., Markides K.E., Lee M.L.
    POLYSILOXANES CONTAINING THERMALLY STABLE CHIRAL AMIDE SIDE-CHAINS
    FOR CAPILLARY GAS AND SUPERCRITICAL FLUID CHROMATOGRAPHY.
    J. Chromatogr., 405, 169-77, (1987).

2. Hara S., Dobashi A., Kinoshita K., Hondo T., Saito M., Senda M.
    CARBON DIOXIDE SUPERCRITICAL FLUID CHROMATOGRAPHY ON A CHIRAL
    DIAMIDE STATIONARY PHASE FOR THE RESOLUTION OF D- AND L-AMINO ACID
    DERIVATIVES.
    J. Chromatogr., 371, 153-8, (1986).

3. Roeder W., Ruffing F.J., Schomburg G., Pirkle W.H.
    CHIRAL SFC-SEPARATIONS USING POLYMER-COATED TUBULAR FUSED SILICA
    COLUMNS. COMPARISON OF ENANTIOMERIC SELECTIVITY IN SFC AND LC USING
    THE SAME STATIONARY PHASE OF THE PIRKLE TYPE.
    J. High Resolut. Chromatogr. Chromatogr. Commun., 10, 665-7, (1987).

CAPILLARY ZONE ELECTROPHORESIS (CZE)

1.   Gozel P., Gassmann E., Michelsen H., Zare R.N.
     ELECTROKINETIC RESOLUTION OF AMINO ACID ENANTIOMERS WITH COPPER (II)-
     ASPARTAMINE SUPPORT ELECTROLYTE
     Anal. Chem.,m 59, 44-49, (1987).

AUTHORS TO PAPERS CROSS-REFERENCE

Abidi S.L.	HPLC/LC 1
Aburatani R.	HPLC/LC 58, 59, 60
Adzuma K.	TLC 18
Aggarwal S.K.	GC 2, SFC 1
Agranat I.	HPLC/LC 35
Alak A.	TLC 1
Alembik M.C.	HPLC/LC 2, 93, 94
Alessi D.M.	HPLC/LC 72
Alexander G.	GC 1
Ali I.	TLC 3, 4
Allenmark S.G.	GEN REV 1
Allmendinger H.	GC 16, 17, 18
Altemura P.	HPLC/LC 71
Arm H.	HPLC/LC 11
Armstrong D.W.	HPLC/LC 3, 4, 29, 78, TLC 1, 2
Ardvidsson T.	HPLC/LC 67
Avgerinos A.	HPLC/LC 5
Ballasteros P.	HPLC/LC 6
Barnes S.	HPLC/LC 30
Bayer E.	GC 18, 19
Beck O.	GC 3
Benschop H.P.	GC 6
Bertucci C.	HPLC/LC 7
Bhushan R.	TLC 3, 4
Blaschke G.	GEN REV 3
Blessington B.	HPLC/LC 9
Bopp R.J.	GEN REV 2
Bradshaw J.S.	GC 2, SFC 1
Brinkman U.A.Th.	TLC 5
Bruner C.A.	TLC 15
Beneken-Kolmer E.W.J.	HPLC/LC 98
Bergos A.	HPLC/LC 87
Berndt H.	HPLC/LC 40
Berni R.J.	HPLC/LC 30
Bertucci C.	HPLC/LC 7, 71
Blaschke G.	HPLC/LC 8
Blessington B.	HPLC/LC 9
Bojarski J.	HPLC/LC 85, 108, 109
Bopp R.J.	HPLC/LC 10
Bosman J.	HPLC/LC 22
Bosshardt H.P.	HPLC/LC 52
Bottorff M.B.	HPLC/LC 42
Bradshaw J.S.	GC 2, SFC 1
Bruegger R.R.	HPLC/LC 11

Buckley D.G.	HPLC/LC 97
Bystricky S.	HPLC/LC 84
Caldwell J.	HPLC/LC 12, 32
Carunchio V.	HPLC/LC 13
Caude M.	HPLC/LC 45, 87
Chemlal A.	HPLC/LC 6
Claramunt R.M.	HPLC/LC 6
Clark B.J.	HPLC/LC 54
Cohen A.I.	GC 14
Coors C.	HPLC/LC 14
Corradini C.	HPLC/LC 15
Coutts R.T.	GC 4, 25
Crabb M.	HPLC/LC 9
Darbyshire J.F.	HPLC/LC 12
Darden T.	HPLC/LC 46
Davies B.E.	GC 5
Degenhardt C.E.A.M.	GC 6
De Jong L.P.A.	GC 6
Delee E.	HPLC/LC 44
Demeter D.A.	HPLC/LC 46
De Zeeuw R.A.	HPLC/LC 22, 23
Dobashi A.	HPLC/LC 16, 17, SFC 2
Dobashi Y.	HPLC/LC 16, 17, 18
Doyle T.D.	TLC 15
Drauz K.	TLC 10
Drenth B.F.H.	HPLC/LC 22, 23
Edholm L.A.	HPLC/LC 19
Einarsson S.	HPLC/LC 20
Eliaz M.	HPLC/LC 35
Elguero J.	HPLC/LC 6
Erlandsson P.	HPLC/LC 21, PREP 1
Ernst-Cabrera K.	GC 7
Fanali S.	HPLC/LC 13, 81
Faulkner J.R.	TLC 2
Faull K.F.	GC 3
Federici F.	HPLC/LC 15, 81, 82
Feitsma K.G.	HPLC/LC 22, 23
Feldberg R.S.	TLC 6
Fell A.F.	HPLC/LC 54
Folestad S.	HPLC/LC 20
Forgione A.	HPLC/LC 65
Finn J.M.	HPLC/LC 24
Fraenkel W.	GEN REV 3
Frank H.	GC 8
Fu P.P.	HPLC/LC 103, 104
Fujimoto Y.	HPLC/LC 25
Fukumoto T.	HPLC/LC 58
Gaget C.	GC 8, 26, 27, HPLC/LC 89, 90
Gallego-Preciado M.	HPLC/LC 6
Gargaro G.	HPLC/LC 26
Gasparrini F.	HPLC/LC 26
Gassman E.	CZE 1
Gazdag M.	HPLC/LC 27
Gioeli C.	HPLC/LC 68
Gont L.K.	TLC 7
Gozel P.	CZE 1

Gray A.	GC 10
Grinberg N.	TLC 8
Groetzinger J.	HPLC/LC 40
Guebitz G.	HPLC/LC 28
Gunther K.	TLC 9, 10, 11, 12
Gyllenhaal O.	GC 11
Hagen M.	GC 21
Halterman R.L.	GC 12
Han S.M.	HPLC/LC 3, 4, 29, TLC 2
Han Y.I.	HPLC/LC 3, 29
Hansson L.	HPLC/LC 21, PREP 1
Hara S.	HPLC/LC 16, 17, 18, GC 13, SFC 2
Hasegawa M.	HPLC/LC 105, 106
Hashima E.	HPLC 64
Hashimoto H.	HPLC/LC 101
Hatada K.	HPLC/LC 58, 59, 60, 61, 62, 63, 64
Hattori T.	HPLC/LC 51
Hayes S.M.	HPLC/LC 30
Heintzelmann B.	GC 9, GC 26, GC 27, HPLC/LC 89, 90
Hoerr D.C.	HPLC/LC 96
Ho M.H.	HPLC/LC 30
Hondo T.	SFC 2
Hoong L.K.	GC 12
Hussenius A.	HPLC/LC 31
Huszar L.	HPLC/LC 27
Hutt A.J.	HPLC/LC 5, 12, 32
Hyun M.H.	HPLC/LC 33, 72
Ichida A.	HPLC/LC 34
Ichikawa M.	HPLC/LC 51
Igwemezie L.	HPLC/LC 49
Irie M.	HPLC/LC 63
Isaksson R.	HPLC/LC 21, 31, 35, PREP 1
Ishi K.	HPLC/LC 25
Ishikura M.	HPLC/LC 64
Israely Z.	HPLC/LC 35
Itabashi Y.	HPLC/LC 86
Iwaki K.	HPLC/LC 36, 37, 53
Jamali F.	GC 4, 25
Jemal M.	GC 14
Johns D.	HPLC/LC 38
Jonas G.	HPLC/LC 97
Josefsson B.	HPLC.LC 20
Juffmann F.	HPLC.LC 28
Jullien I.	HPLC/LC 44
Jurczak J.	GC 20
Juvancz Z.	GC 1
Kaida Y.	HPLC/LC 60
Kakimoto T.	HPLC/LC 25
Kameyama K.	TLC 18
Kamminga D.	TLC 5
Karlsson A.L.	HPLC/LC 67
Karnes H.T.	HPLC/LC 39
Kawashima M.	HPLC/LC 62
Kayano M.	HPLC/LC 51
Kennedy J.H.	GEN REV 2, HPLC/LC 10
Kerr C.R.	HPLC/LC 49
Kimoto H.	HPLC/LC 105

Kinkel J.N.	GEN REV 3
Kinoshita K.	HPLC/LC 17, 36, 37, 53, SFC 2
Kita H.	HPLC/LC 101
Kitahara H.	HPLC/LC 56
Kobayashi M.	HPLC/LC 79
Koch E.M.	GC 19
Koenig W.A.	GC 7, 15, 21, GEN REV 4
Kooi K.H.	HPLC/LC 22
Koppenhoefer B.	GC 16, 17, 18, 19
Koscielski T.	GC 20
Kraak J.C.	HPLC/LC 88
Krueger G.	HPLC/LC 40
Laarhoven W.H.	HPLC/LC 75
Ladanyi L.	HPLC/LC 41, TLC 14
Lalonde R.L.	HPLC/LC 42, 83
Lam S.	HPLC/LC 43
Lamm B.	GC 11
Lee M.L.	GC 2, SFC 1
Le Garrec L.	HPLC/LC 44
Legendre M.G.	HPLC/LC 30
Lienne M.	HPLC/LC 45, 87
Lindberg C.	HPLC/LC 19
Ling D.	HPLC/LC 102
Link R.	GC 24
Lipkowitz K.B.	HPLC/LC 46
Liu R.H.	HPLC/LC 30
Lu X.L.	HPLC/LC 47
Lutz S.	GC 21
Malikin G.	HPLC/LC 43
Markides K.E.	GC 2, SFC 1
Marle I.	HPLC/LC 67
Martens J.	TLC 9, 10, 11
Marti A.R.	HPLC/LC 11
Mathews B.L.	GC 10
Matsson O.	HPLC/LC 31
Matusch R.	HPLC/LC 14
McDaniel D.M.	HPLC/LC 48
McErlane K.M.	HPLC/LC 49
Mehta A.C.	HPLC/LC 50
Meijer D.K.F.	HPLC/LC 23
Messina A.	HPLC/LC 13, 15, 81, 82
Meyer V.R.	HPLC/LC 11, PREP 2
Michelson H.	CZE 1
Mihashi K.	GC 13
Minarik M.	HPLC/LC 99, 100
Misiti D.	HPLC/LC 26
Miwa T.	HPLC/LC 51
Miyakawa T.	HPLC/LC 51
Miyake Y.	HPLC/LC 51
Miyano S.	HPLC/LC 101
Mueller M.D.	HPLC/LC 52
Mushtaq M.	HPLC/LC 103, 104
Mutschler E.	HPLC/LC 70
Neuendorf S.K.	TLC 7
Nicholson G.J.	GC 19
Nimura N.	HPLC/LC 36, 37, 53
Nishi H.	HPLC/LC 25
Nishimura K.	HPLC/LC 80

Noctor T.A.G.	HPLC/LC 54
Norinder U.	HPLC/LC 55
Numakura T.	HPLC/LC 101
Ogura H.	HPLC/LC 36, 37
Oi N.	HPLC/LC 56
Oi S.	HPLC/LC 101
Okabe H.	GC 13
Okamoto Y.	GEN REV 4, HPLC/LC 57, 58, 59, 60, 61, 62, 63, 64
Ossig A.	GC 24
O'Sullivan J.	HPLC/LC 9
Oya S.	DCC 1
Palmieri G.	HPLC/LC 26
Parish C.A.	HPLC/LC 46
Pascal J.C.	HPLC/LC 44
Pasutto F.M.	GC 4, 25
Paulson J.	HPLC/LC 19
Pedrazzini S.	HPLC/LC 65
Perry J.A.	PREP 3
Pettersson C.	HPLC/LC 66, 67, 68, 69
Pflugman G.	HPLC/LC 70
Pierini M.	HPLC/LC 26
Pillion D.J.	HPLC/LC 30
Pini D.	HPLC/LC 7, 71
Pirkle W.H.	HPLC/LC 33, 72, 73, 74, SFC 3
Pochapsky T.C.	HPLC/LC 72, 73
Prinsen W.J.C.	HPLC/LC 75
Railton I.D.	HPLC/LC 76
Rateike J.D.	PREP 3
Rausch R.	TLC 12, 13
Ravichandran K.	HPLC/LC 77
Repke D.B.	GC 3
Repucci L.M.	TLC 6
Rieck M.	GC 21
Reid E.	GEN REV 5
Roeder W.	SFC 3
Rogers L.B.	HPLC/LC 77
Rosini C.	HPLC/LC 7, 71
Rouse C.A.	GC 2, SFC 1
Roush W.R.	GC 12
Roussel C.	HPLC/LC 6
Rosset R.	HPLC/LC 45, 87
Ruffing F.J.	SFC 3
Sacchi S.	HPLC/LC 65
Saigo K.	HPLC/LC 105, 106
Saito M.	SFC 2
Sakamoto H.	HPLC/LC 63
Salvadori P.	HPLC/LC 7, 71
Sandstroem J.	HPLC/LC 35
Sarkar M.A.	HPLC/LC 39
Scales B.	GEN REV 5
Schickedanz M.	TLC 9, 10, 11
Schill G.	HPLC/LC 69
Schomburg G.	SFC 3
Schurig V.	GC 22, 23, 24, PREP 4
Self C.	GC 10
Secor H.V.	HPLC/LC 4, 78
Seeman J.I.	HPLC/LC 4, 78

Senda M.	SFC 2
Shibata T.	HPLC/LC 34, 95
Shimada A.	TLC 18
Shimizu R.	HPLC/LC 25
Shimizu T.	HPLC/LC 79
Shinbo T.	HPLC/LC 80
Siekerman C.	HPLC/LC 88
Singh N.N.	GC 25
Singh N.	GC 4
Sinibalda M.	HPLC/LC 13, 15, 81, 82
Siegel P.	HPLC/LC 41, TLC 14
Smith E.	HPLC/LC 94
Snider B.G.	HPLC/LC 48
Snyder J.K.	DCC 1
Sowin T.J.	HPLC/LC 74
Spahn H.	HPLC/LC 70
Spino L.A.	HPLC/LC 4
Stiffin R.M.	HPLC/LC 95
Straka R.J.	HPLC/LC 83
Strasak M.	HPLC/LC 84
Sugiura M.	HPLC/LC 80
Sun Z.	HPLC/LC 102
Sundholm E.G.	HPLC/LC 55
Suzuki T.	HPLC/LC 101
Sybilska D.	GC 20, HPLC 85, 108, 109
Szczerba T.J.	PREP 3
Szejtli J.	GC 1, HPLC/LC 109
Szepesi G.	HPLC/LC 27
Sztruhar I.	HPLC/LC 41
Tachibana K.	HPLC/LC 105, 106
Takagi T.	HPLC/LC 86
Takai N.	HPLC/LC 101
Takeda K.	HPLC/LC 36, 37
Tambute A.	HPLC/LC 45, 87
Tarbet B.J.	GC 2, SFC 1
Timmons K.D.	HPLC/LC 78
Toth-Lauritz M.	TLC 14
Tsang W.S.	HPLC/LC 30
Tsumagari N.	HPLC/LC 25
Tsuno M.	HPLC/LC 51
Van Beijsterveldt L.E.C.	HPLC/LC 98
Van de Meent D.	GC 6
Van den Berg G.R.	GC 6
Van der Kleyn E.	HPLC/LC 98
Van Genderen J.	GC 6
Vereczekey-Donath G.	HPLC/LC 41, TLC 14
Vessman J.	GC 11
Villani D.	HPLC/LC 26
Vloon W.J.	HPLC/LC 88
Wagner J.	GC 9, 26, 27, HPLC/LC 89, 90
Wainer I.W.	HPLC/LC 2, 42, 83, 91, 92, 93, 94, 95, TLC 15, GEN REV 6, 7
Walhagen A.	HPLC/LC 19
Ward T.J.	HPLC/LC 78
Weaner L.E.	HPLC/LC 96
Weems H.B.	HPLC/LC 104
Weinstein S.	TLC 8, 16
Whelpton R.	HPLC/LC 97
Wilson I.D.	TLC 17, GEN REV 5

Winter S.M.	HPLC/LC 12
Wolf E.	GC 9, 26, 27, HPLC/LC 89, 90
Wuis E.W.	HPLC/LC 98
Wulff G.	HPLC/LC 99, 100
Yamaguchi T.	HPLC/LC 80
Yamashita J.	HPLC/LC 101
Yang Q.	HPLC/LC 102
Yang S.K.	HPLC/LC 47, 103, 104
Yasui M.	TLC 18
Yoshida S.	HPLC/LC 36, 37
Yuasa S.	TLC 18
Yuki Y.	HPLC/LC 105, 106
Zanoboni-Muciaccia W.	HPLC/LC 65
Zare R.N.	CZE 1
Zief M.	HPLC/LC 107
Zukowski J.	HPLC/LC 85, 108, 109

Abscisic acid methyl ester. HPCL/LC 76
Acids. HPLC/LC 68
Acyloins. GC 21
α-alanine. HPLC/LC 90
β-alanine. HPLC/LC 90
α-alkyl amino acids. TLC 10
Aldose. GC 13
Alkyl aryl carbinols. HPLC/LC 71
Alkyl-substituted indene enantiomers. HPLC/LC 31
Amides. HPLC/LC 94
Amines. HPLC/LC 43
Amino acid derivatives. HPLC/LC 40, SFC 2
Amino acid enantiomers, HPLC/LC 20
Amino acids. HPLC/LC 13, 80, 106, TLC 3, CZE 1
β-amino alcohols. HPLC/LC 102
DNS-amino. TLC 8
γ–aminobutyric acid. HPLC/LC 90
Amino compounds. HPLC/LC 36
Amphetamine. HPLC/LC 2, 30
Anomeric purine nucleosides. TLC 6
Anticholinergic drugs, HPLC/LC 22
Aromatic alcohols. HPLC/LC 95
α-Arylacetic acids. HPLC/LC 48
2-Arylpropionic acids. GC 25
Baclofen. HPLC/LC 98
Barbiturates. HPLC/LC 85, 108, 109
Cadaverine analogs. GC 9
N-carbamyltryptophan. TLC 7
Carboxylic acid. DCC 1, HPLC/LC 60
Carboxylic acids (as diastereomeric carboxamides). TLC 14
Carboxylic acids (by derivatization). HPLC/LC 25, 41
2-(4-Chloro-2-methylphenoxy) propanoic acid. HPLC/LC 9
Cobalt (III) chelate complexes. HPLC/LC 84
Cromakalim. GC 5
Cyanohydrins. GC 21
Dialkylglycerol. HPLC/LC 86
Diacylglycerol. HPLC/LC 86
Dihydrodiol enantiomers of benz[A]anthracene. HPLC/LC 103
3,5'-Dimethyl-4,4'-dibromo-1,1'-bispyrazolylphenylmethane. HPLC/LC 6
Diol enantiomers of polycyclic aromatic hydrocarbons. HPLC/LC 104
1,6-Dioxaspiro[4.4]nonanes. GC 23, PREP 4
Dipeptides. GC 18, HPLC/LC 72
Dopamine enantiomers. HPLC/LC 53
Dopamine derivatives. HPLC/LC 53
Ethyl N,N-dimethylphosphoramidocyanidate (Tabun). GC 6
Fatty acid amide, HPLC/LC 96

Fatty acid ester, HPLC/LC 96
Flamprop M-isopropyl. GC 10
Flavors. GC 22
Flunoxaprofen. HPLC. LC 65
Halo carboxylic acid amides. GC 19
Halo carboxylic acid esters. GC 24
Hexahelicene (and methyl substituted derivatives), HPLC/LC 75
Hydroxy acids. GC 16
6-Hydroxymethtryptoline. GC 3
Ibuprofen. HPLC/LC 5
Imino alcohol derivatives of chiral phthalides. HPLC/LC 74
Mephenytoin. HPLC/LC 85
Metamphetamine. HPLC/LC 30
Metoprolol. HPLC/LC 70
Mexilethine. HPLC/LC 49
Nefopam hydrochloride. HPLC/LC 35
Nicotine. HPLC/LC 4, 78
Nicotine analogs. HPLC/LC 4, 78
Non-steroidal anti-inflammatory drugs. GC 4
Norepinephrine. HPLC/LC 53
Oxyphenonium bromide. HPLC/LC 23
2-phenylpropionic acid NSAID's. HPLC/LC 32
Phenylthiohydantoin amino acids. TLC 4
Pheromones. GC 22
Phthalides. HPLC/LC 74
Primary alcohols. GC 17
L-proline benzyl ester. GC 14
Propranolol. HPLC/LC 83
Propionic acid-derived herbicides. HPLC/LC 52
Rotenoids. HPLC/LC 1
Secondary alcohols. GC 17
Secondary methyl ethers. GC 12
Selenoxides. HPLC/LC 26, 79
D- & L-Solatol Mandelate. HPLC/LC 44
Substituted putrescine. GC 9
Sulfoxides. HPLC/LC 26
Taban, GC 6
Tertiary alcohols. GC 17
Thioridazine (also metabolites & related compounds). HPLC/LC 97
Thyroid hormones. HPLC/LC 28
Tocainide (and some of its analogs). GC 11
2,2,2-Trifluoro-1-(9-anthyrl) ethanol. HPLC/LC 46
Tripeptides. HPLC/LC 72
DL-Tryptophan. TLC 18

SUBJECT REFERENCE - CHIRAL STATIONARY PHASES & ELUENTS

Albumin in the mobile phase. HPLC/LC 67
Albumin adsorbed to silica. HPLC/LC 21, PREP 1
Amide enantiomers. HPLC/LC 96
Amine derivative-bonded silica gel. HPLC/LC 82
Amylose tris(3,5-dimethylphenylcarbamate). HPLC/LC 58, 60
Berberine-impregnated silica (for amino acids). TLC 3
Binaphthalene derivatives bonded to silica gel. HPLC/LC 101
"Brush type" chromatographic stationary phases. HPLC/LC 11
S-camphor-10-sulphonyl derivatives (for drugs). HPLC/LC 14
Cellulose derivatives. HPLC/LC 34
Cellulose (for DL-tryptophan). TLC 18

Cellulose trans-tris (4-phenylazophenylcarbamate). HPLC/LC 63
Cellulose tribenzoate (for amides). HPLC/LC 94
Cellulose tribenzoate (for aromatic alcohols). HPLC/LC 95
Cellulose tribenzoate derivatives. HPLC/LC 59
Cellulose tris (3,5-dimethylphenylcarbamate). HPLC/LC 60
Cellulose tris(3,5-dimethylphenylcarbamate)-coated silica gel (for abscisic acid methyl ester). HPLC/LC 76
Cellulose tris(3,5-dimethylphenylcarbamate) (for propranolol). HPLC/LC 83
Chelate complexes with copper(II)-L-proline (for amines). HPLC/LC 43
Chiral additives (in the mobile phase). HPLC/LC 66, 81
Chiral complexation gas chromatography (for secondary methyl ethers). GC 12
Chiral diamide phases. HPLC/LC 17
Chiral diamide stationary phase (for D- & L-amino acid derivatives). SFC 2
Chiral diamine-bonded silica gel. HPLC/LC 15
Chirasil-val (for dipeptides). GC 18
Chirasil-val (for Halo carboxylic acid amides). GC 19
Chirasil-val (for tocainide & analogs). GC 11
Cis-tris(4-phenylazophenylcarbamate). HPLC/LC 63
Complexation gas chromatography. GC 6
Copper(II)aspartamine, CZE 1
Crown ether-coated reversed-phase packings (for amino acids). HPLC/LC 80
α-cyclodextrin. GC 20, HPLC/LC 108
β-cyclodextrin. HPLC/LC 108
β-cyclodextrin-bonded phases (for aromatics). HPLC/LC 3
β-cyclodextrin (for nicotine & analogs). HPLC/LC 78
β-cyclodextrin inclusion complexes (for mephenytoin & barbiturates). HPLC/LC 85
β-cyclodextrin inclusion complexes (for nicotine & analogs). HPLC/LC 4
α-, β- & γ-cyclodextrins (as mobile phase additives). HPLC/LC 27
Cyclodextrins (and their derivatives). GC 1
L-N-(3,5-dimethoxybenzoyl)isoleucine. HPLC/LC 88
Dipeptides for Zwitter-ion pair chromatography (for amino acids). HPLC/LC 77
Heptakis (2,6-di-O-methyl)-β-cyclodextrin. HPLC/LC 108
Hydroxyethyl-derivatized β-cyclodextrins. TLC 2
Hydroxypropyl-derivatized β-cyclodextrins. TLC 2
Immobolized acetylquinine (for acids). HPLC/LC 68
Ion-pairing. HPLC/LC 69
Ligand-exchange. HPLC/LC 105, TLC 9, 12
Ligand-exchange chromatography. HPLC/LC 15, 106
Ligand-exchange chromatography (for thyroid hormones). HPLC/LC 28
Ligand-exchange chromatography (of amino acids). HPLC/LC 13
Methylated β-cyclodextrins (for barbiturates). HPLC/LC 109
Microcrystalline triacetylcellulose (for indene enantiomers). HPLC/LC 31
Ovomucoid as a column ligand. HPLC/LC 51
Phosphorous atom as assymetric center. HPLC/LC 87
Pirkle-type chiral stationary phases. HPLC/LC 24
Polyacrylamides as chiral phases (for drugs). HPLC/LC 8
Polymeric sorbents. HPLC/LC 99
Polysiloxanes. GC 7
Polysiloxanes containing thermally stable chiral amide side-chains. GC 2, SFC 1
Poly(triphenylmethyl methacrylate). HPLC/LC 61
Poly(triphenylmethyl methacrylate) derivatives. HPLC/LC 64
Silica-supported quinine (for alkyl aryl carbinols). HPLC/LC 71
(+)-Tartaric acid-impregnated silica gel (for phenylthiohydantoin). TLC 4
(R,R)-Tartramide (derived CSP). HPLC/LC 18
Triacetylcellulose (for 3,5'-dimethyl-4,4'-dibromo-1,1'-bispyrazolylphenylmethane). HPLC/LC 6
Tripeptides for Zwitter-ion pair chromatography (for amino acids). HPLC/LC 77
Tris(3,5-dichlorophenylcarbamate) on silica gel. HPLC/LC 58
Urea derivatives of L-valine. HPLC/LC 56

SUBJECT REFERENCE - DERIVATIZATION REACTIONS

Activated carbamate (as chiral derivatizing agent for amino compounds). HPLC/LC 36
(+)- or (-)-Amphetamine as derivatizing reagent (for 2-arylpropionic acids). GC 25
Derivatization (for carboxylic acids). HPLC/LC 41
Derivatization reactions for hydroxy acids. GC 16
Derivatization reagents (for the resolution of carboxylic acid enantiomers). HPLC/LC 25
Derivatization with O-phthalaldehyde and 2,3,4,6-tetra-O-acetyl-1-thio-β-glucopyranoside.
    HPLC/LC 20
Hexahelicene (and methyl substituted derivatives). HPLC/LC 75
(S)-(-)-phenylethyl isocyanate (for metoprolol). HPLC/LC 70
New derivatisation reagents for carboxylic acids, HPLC/LC 25
O-Phthalaldehyde (for baclofen). HPLC/LC 98
O-Phthalaldehyde-N-acetyl-L-cysteine (derivatization for norepinephrine enantiomers).
    HPLC/LC 53
Trimethylsilyl ethers of methyl 2-(polyhydroxyalkyl)thiazolidine-4(R)-carboxylates for
    aldose, GC 13
Urea derivatives (for β-amino alcohols). HPLC/LC 102

SUBECT REFERENCE - DETECTORS

Capillary gas chromatography and high resolution mass spectrometry. GC 3
Chromatography-mass spectrometry (for amphetamine & metamphetamine). HPLC/LC 30
Circular dichroism detection. HPLC/LC 7
Circular dichroism spectroscopy. HPLC/LC 35
Fluorescence detection. HPLC/LC 49
Liquid chromatography-mass spectrometry. HPLC/LC 19

SUBJECT REFERENCE - DRUGS/PHARMACEUTICALS

Analysis of chiral drugs. HPLC/LC 12
Anticholinergic drugs, HPLC/LC 22
Cardiovascular drugs. HPCL/LC
Drug compounds. HPLC/LC 39
Drug enantiomers. HPLC/LC 50, 54
Drug enantiomers in biological samples. HPLC/LC 19
Drugs & metabolites. GC 8, HPLC/LC 23
Enantiomeric drugs. HPLC/LC 102
Liquid chromatography of drugs. HPLC/LC 14
Non steroidal anti inflammatory drugs. GC 4
Pharmaceutical compounds. GEN REV 2
Resolution of drugs. HPLC/LC 8
Stereochemical purity of drugs. HPLC/LC 7

APPENDIX 2 — SOME MANUFACTURERS AND SUPPLIERS OF CHIRAL COLUMNS

Alltech Associates
6-7 Kellet Road Industrial Estate
Carnforth
Lancashire
LA5 9XP
UK

Anachem
20 Charles Street
Luton
Bedfordshire
LU2 OEB
UK

Applied Science
PO Box 440
State College
PA 16801
USA   .

J T Baker
222 Red School Lane
Phillipsburg
NJ 08865
USA

Daicel Chemical Industries
Konigsallee 92a
D-4000 Dusseldorf 1
FRG

Daicel Chemical Industries
New York
NY 10166-0130
USA

Hichrom Ltd
6 Chiltern Enterprise Centre
Station Road
Theale
Reading
Berkshire
RG7 4AA
UK

Alltech Associates Inc
2051 Waukengan Road
Deerfield
IL 60015
USA

AN-Anspec
Ann Arbor
MI 48107
USA

ASTEC
37 Leslie Court
PO Box 297
Whippany
NJ 07981
USA

J T Baker Chemicals BV
PO Box 1
7400 AA Deventer
The Netherlands

Daicel Chemical Industries
8-1 Kasumigaseki 3-chome
Chiyoda-ku
Tokyo 100
Japan

Field Analytical
PO Box 113
Weybridge
Surrey
KT13 8BJ
UK

Jones Chromatography
Tir-y-Berth Industrial Estate
New Road
Hengoed
Mid Glamorgan
UK

Macherey-Nagel GmbH
PO Box 307
D-5160 Durren
FRG

Perstorp Biochem Biotechnology
Ideon
S-22370 Lund
Sweden

Phase Separations
Deeside Industrial Park
Deeside
Clwyd
CH5 2NU
UK

Serva Feinbiochemica
Postfach 105260
Kitahama
D-6900 Heidelberg 1
FRG

Supelco Inc
Supelco Park
Bellefonte
PA 16823
USA

Technicol
Brook Street
Higher Hillgate
Stockport
Cheshire
SK1 3HS
UK

E. Merck
Frankfurter Strasse 250
D-6100 Darmstadt
FRG

Pharmacia LKB Biotechnology
Box 308
S-1626 Bromma
Sweden

Regis Chemical Co
8260 Austin Avenue
Morton Grove
IL 60053
USA

Sumitomo Chemical Co
Sumitomo Building 5-15 Kitahama
Higashi-ku
Osaka 541
Japan

Supelchem
London Road
Sawbridgeworth
Hertfordshire
CM21 9JH
UK

Y - YMC Inc
ME Freedom
NJ 07090
USA

APPENDIX 3 — ABSTRACTS

The abstracts provided here are for those papers and posters for which no manuscript was received for publication. They have been retyped but are otherwise printed as received.

STRATEGIES FOR RESOLVING ENANTIOMERS OF VARIOUS α-SUBSTITUTED AMINOACIDS: COMPARISON OF HPLC AND GC PROCEDURES (PAPER)

J. Wagner, E. Wolf, B. Heintzelmann, M. Zreika
and M.G. Palfreyman

Merrell Dow Research Institute
16 rue d'Ankara, 67084 Strasbourg-Cedex, France

The separations of enantiomers of aminoacids by chromatographic methods are all based on the differences in properties of the two enantiomers in an asymmetric surroundings. We have applied the different methods available, i.e. GC with chiral phases or GC with achiral phases (after derivatization into diastereoisomers) for the resolution of various substituted aminoacids. HPLC separations on achiral phases after derivatization with chiral reagents were also explored. Our main effort was devoted towards the application of ligand exchange chromatography[1] for the resolution of the underivatized aminoacids. The use of chiral phases with copper salts in the eluent or achiral phases with a chiral eluent, i.e. (L)-proline or N,N-di-n-propyl-(L)-alanine/copper complex [2,3,4], allowed the resolution of nearly all the aminoacids studied. Various substituted analogues of glutamic acid, ornithine, lysine, arginine, phenylalanine and tyrosine were resolved. The advantages and limitations of the two ligand-exchange procedures, chiral phase or chiral eluent, will be discussed. The chiral eluent reversed phase HPLC procedure has been applied to the semipreparative scale resolution of the enantiomers of (E)-β-fluoro-methylene-m-tyrosine (MDL 72394), a substrate for aromatic-L-aminoacid decarboxylase and prodrug of a potent monoamine oxidase inhibitor. Data about the biochemical activities of the (D)- and (L)-enantiomers will be presented.

REFERENCES

1. V. A. Davankov, A. A. Kurganov and A. S. Bochkov, *Adv. Chromatogr.* (NY) 71:136 (1983).
2. E. Gil-Av, A. Tishbee and P. E. Hare, *J.Am.Chem.Soc.* 102:5115 (1980).
3. S. Weinstein, *Agnew.Chem.Int.Ed.Engl.* 21:218 (1982).
4. J. Wagner, E. Wolf, B. Heintzelmann and C. Gatet, *J.Chromatogr.* 391:211 (1987).

BONDED CYCLODEXTRINS FOR CHIRAL SEPARATIONS UTILIZING INCLUSION COMPLEX FORMATION (PAPER)

T. E. Beesley

Advanced Separation Technologies Inc
37 Leslie Court, PO Box 297
Whippany, NJ 07981, USA

By virtue of their structure, cyclodextrins create a reversible, encapsulating environment in which various organic and inorganic molecules can enter and orientate themselves allowing highly specific interactions. The fundamental mechanism is referred to as inclusion complexing, a term coined by Professor Freundenberg in 1938. It involves the interaction of the hydrophobic cavity of the cyclodextrin toroidal structure and the lipophilic portions of an analyte. The cyclodextrin ring, formed by 1,4 glycoside linkage of glucose molecules, has 5 chiral centers per molecule giving beta-cyclodextrin 35 chiral sites in total, and optimizing the potential for chiral recognition. Other mechanisms in this technique involve the primary and secondary hydroxyls positioned at the opening of the cavity.

It it now known that these secondary mechanisms characterize potential separation. Computer graphics of crystal data show the orientation of the secondary hydroxyls to be in an opposite direction to the primary hydoxyls and the interaction of the solute with these will define potential separation. In addition, they can be reacted with various groups to effect selectivity changes. It is also known that for chiral recognition the analyte needs to fit the cavity as closely as possible, and the degree of fit therefore provides a variable binding strength. Increasing ring size or ring number as well as adding substituents to either situation can affect retention behavior. Of the three available cyclodextrins, alpha cyclodextrin shows one half of the affinity of molecules of the size of benzene compared to beta cyclodextrin, which shows one half the affinity for pyrene as compared to gamma cyclodextrin. Beta cyclodextrin therefore accommodates structures from benzene to naphthalene and gamma naphthalene to chrysene. Substituents in any of these ring structures affect their ability to include and consequently alters retention behavior. Hydrogen bonding interactions with secondary hydroxyl groups of the cyclodextrin will also affect retention behavior, while solvent interaction plays an important role.

Since the introduction of silica bonded cyclodextrins as an HPLC stationary phase in 1984, many applications have been generated that elaborate the potential of this technique in chiral separations. Computer analysis of crystal data has recently been extensively used to predict and explain interations. For the separation of d- and l- propanolol, computer projections illustrate the optimal orientation of each isomer on the basis of the highest degree of hydrogen bonding and complexation. Bond lengths can be measured from the data, successfully predicting Van der Waals interactions with the secondary hydroxyls.

Method development on bonded cyclodextrins is similar to that used for reversed phase stationary phases. Acetonitrile and ethanol exhibit a greater affinity for the cyclodextrin cavity than methanol and consequently can be used as modifiers to alter selectivity. The use of different solvent combinations, including buffers, and the effects of flow rate and temperature are illustrated by the optimization of enantiomeric separation of barbiturates, steroids, the antihistamine chlorpheniramine, and peptides and dansyl amino acids. Applications for cyclodextrin bonded phases are increasingly also being found in the fungicide and pesticide industry, as exemplified by the chiral separation of triadimenol enantiomers.

APPLICATIONS OF COMPUTERIZED MOLECULAR GRAPHICS TO ENANTIOSEPARATION IN HPLC (PAPER)

D. R Taylor, G. Bridger, R. Husband* and M. J. Tandy*

Chemistry Department, UMIST, Manchester M60 1QD,
UK and *ICI Plant Protection, Jealotts Hill Research
Station, Bracknell, Berkshire RG12 6EY, UK

An Evans & Sutherland computerized molecular graphics (CMG) system has been used to establish the optimal conformations of enantiomeric analytes and the organic portion of covalently bonded chiral HPLC stationary phases. The CMG workstation also enables the association complexes between the chiral stationary phase (CSP) and enantiomeric analytes to be visualized, a necessary step in rationalizing observed elution orders of enantiomers during separation on CSP.

Two typical studies will be described. In the first, a conformationally optimized, minimum energy, conformer of a previously reported[1,2] CSP based on N-formyl-L-phenyl-alanine was associated with a test solute of known elution order and absolute configuration. Structural features which influence the chiral recognition process were identified by use of the CMG workstation and these conclusions were validated experimentally. In particular, separations were confirmed to be improved for these analytes on (a) CSP with longer anchor chains (1) and (b) CSP with re-orientated chiral strands (2).

In the second investigation, a family of new CSP (3) were synthesized from tartaric acid monoanilides and used successfully for the direct HPLC analysis of racemic aminoacid derivatives, aminoalcohols, aryloxpropionates (4) and dihydrothiadiazines (5). The aryloxypropionate separations have also been studied using CMG, and a mechanism will be discussed.

3) 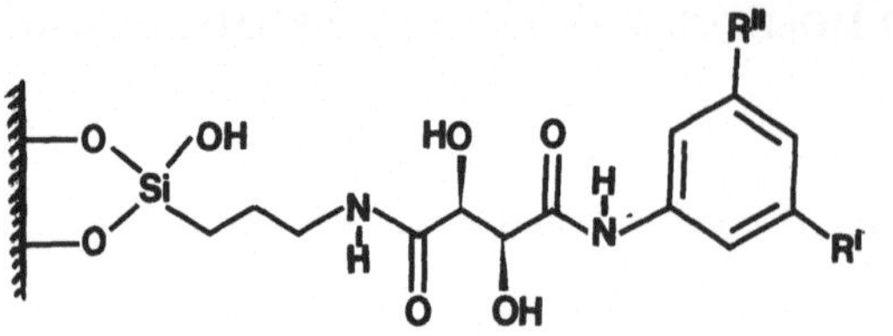

$R^{I}=R^{II}=H$ <u>OR</u> NO_2

$R^{I}=H$, $R^{II}=NO_2$

4) $ArOCHMeCO_2R$

5) 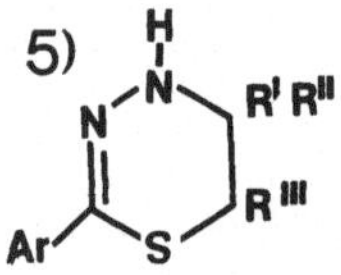

$R^{I}=H$ $R^{II}=Ar$

$R^{III}=CN, CO_2Et, Ph$

For both families of CSP, a study of the relationship between the column packing's surface coverage, column efficiency, and enantioselectivity has enabled optimum performance to be achieved,

REFERENCES

1. J. N. Akanya, D. M. Hitchen, and D. R. Taylor, *Chromatographia*, 16:224 (1982).
2. J. N. Akanya and D. R. Taylor, *J.Liq.Chromatog.* 10:805 (1987).

SEPARATION OF ENANTIOMERS BY GAS CHROMATOGRAPHY: PRINCIPLES AND APPLICATIONS (PAPER)

E. Bayer

Institute for Organic Chemistry
University of Tübingen, FRG

Gas chromatographic separation of chiral metabolites, drugs and natural compounds of various structures can be achieved with Chirasil-Val, a chiral polysiloxane peptide as stationary phase. High sensitivities and fast quantitative analysis are possible. High accuracy in determining enantiomer purity can be obtained, e.g. as low as 0.02% of a "wrong" enantiomer. Besides enantiomer purity of synthetic drugs also enantiomer purity of biotechnologically synthesized drugs and enzymatic enantiomer selectivity has been investigated.

The mechanism of enantiomer discrimination on Chirasil-Val is discussed. In gas chromatography the interaction of selector and selectand can be better investigated because there is no effect of the mobile phase. Thermodynamic data are presented and the molecular mechanism of enantiomer discrimination discussed in detail.

DEVELOPMENT OF STEREOSPECIFIC ASSAY METHODS FOR THE DETERMINATION OF RACEMICALLY ADMINISTERED ANTIARRHYTHMICS (PAPER)

E. Martin, H. Spahn and E. Mutschler

Department of Pharmacology, University of Frankfurt
Frankfurt/Main, FRG

More than two thirds of the antiarrhythmic agents currently used are chiral, the vast majority thereof being administered in racemic form. This situation contrasts with the fact that to time but little is known about the pharmacodynamic and pharmacokinetic properties of the individual enantiomers. Considering both frequency and seriousness of untoward effects displayed by antiarrhythmic agents, the study of stereospecific effects in pharmacokinetics is of particular interest.

Hence it was the aim of the present investigation to develop and compare stereospecific assay methods for monitoring plasma and urine levels of racemically administered antiarrhythmics. Resolution of racemic mixtures was performed by TLC or HPLC, following derivatization with various, optically active amine labels (acid chlorides of benoxaprofen and naproxen, naphthyl-ethyl-isocyanate - NEIC). For this reason only antiarrhythmics with primary and secondary amine functions were selected: flecainide and its meta-O-dealkylated metabolite (MODF), propafenone and diprafenone, mexiletine, tocainide and cibenzoline.

Drugs were either directly extracted from alkaline plasma or urine samples with n-hexane (flecainide, cibenzoline) or in the form of a heptane-sulphonate ion-pair (MODF). Derivatization with NEIC was performed at ambient temperature in a mixture of chloroform/dimethylformamide. Times necessary to achieve maximal derivatization yields varied considerably, ranging from 30' (flecainide) to 16 h (diprafenone). In the case of the acid chlorides (benoxaprofen chloride for mexiletine; naproxen chloride for tocainide) derivatization was performed in toluene at 60°C. Using TLC diastereomeric reaction products were separated on silica gel plates with solvent mixtures containing toluene-dichloromethane-tetrahydrofuran (5+4+1, v/v: flecainide) or toluene-dichloromethane-tetrahydrofuran-methanol (50+37+10+3, v/v: MODF and cibenzoline). Resolution of the derivatives of diprafenone, propafenone, mexiletine, flecainide and MODF by HPLC was performed on a C 18-reversed phase column with solvent mixtures containing methanol and water (diprafenone: 75+25 v/v; mexiletine, flecainide, propafenone: 73+27 v/v; MODF 60+40 v/v).

Detection of the separated diastereomers was performed by monitoring UV-absorption in case of cibenzoline (214nm). All others were detected by measuring the fluorescence intensity, either of the drug itself (flecainide: 313nm/365nm TLC, 281nm/361nm HPLC; MODF: 313nm/365nm TLC, 281nm/375nm HPLC) or of the reagent moiety (mexiletine: 284nm/335nm HPLC; diprafenone and propafenone: 280nm/334nm HPLC).

The applicability of the methods was tested with samples drawn from patients who had received oral or iv doses of the respective drugs.

DIRECT HPLC RESOLUTION OF β-AMINOALCOHOL (TAZIFYLLINE, RANOLAZINE, SOTALOL) ENANTIOMERS (POSTER)

E. Delée, L. Le Garrec, I. Jullien, S. Béranger,
J. C. Pascal and H. Pinhas

Recherche Syntex France, Leuville-Sur-Orge
91319 Montlhéry, France

A routine chiral analysis has been developed to control the optical purity of chiral drugs and to monitor their asymmetric synthesis. The recent advent of new chiral stationary phases for HPLC enabled us to achieve the direct resolution of chiral drugs without any derivatization.

The factors affecting chiral resolution on a new α_1 acid glycoprotein column (EnantioPac, LKB) were assessed with three β-aminoalcohols. Tazifylline (antihistaminic) and Ranolazine (antianginal) are both piperazines with a 2-hydroxypropyl side chain and their two respective enantiomers could be separated with 2-propanol in phosphate buffer pH=7 mobile phase[1]. To improve the resolution, concentration of 2-propanol and flow rate were adjusted. For tazifylline, the data generated yielded an optimum eluent composition of 2-propanol/phosphate buffer pH=7 15/85 v/v. Phosphate buffer composition=4mM NaH_2PO_4+4mM Na_2HPO_4+10mM NaCl. Flow rate =0.25 ml/min, t°=20°C, sample loading = 1 nanomole. For Ranolazine, the data generated yielded an optimum eluent composition of 2-propanol/phosphate buffer pH=7 4/96 v/v. Phosphate buffer composition was the same as tazifylline buffer. Flow rate was 0.20 ml/min, t°=20°C, sample loading = 1 nanomole.

For Sotalol (antiarrhythmic), a β-aminoalcohol with aliphatic secondary amine, the same mobile phase used as above could not give a resolution of two enantiomers. A different organic modifier was necessary and instead of uncharged 2-propanol modifier, an anionic modifier was added in phosphate buffer[2]. To improve the resolution, the nature of anionic acid modifier, its concentration, flow rate of mobile phase, and pH of phosphate buffer were adjusted. Five 0.05M carboxylic acids were assayed (n=4 to 8) and hexanoic acid gave the best resolution with a 0.02M phosphate buffer pH=7 as the mobile phase. The concentration of hexanoic acid was adjusted (0.025M to 0.75M) as well as flow rate of mobile phase (0.15ml/min to 0.25ml/min). We also adjusted the pH (6.4 to 7.2). The data generated yielded the following optimal conditions: mobile phase = 0.05M hexanoic acid in 0.02M phosphate buffer (0.01M NaH_2PO_4+0.01M Na_2HPO_4) pH=7, flow rate = 0.20ml/min, t°=20°C, sample loading = 1 nanomole. There is no improvement of the resolution with NaCl ionic strength adjustment. For all three compounds each enantiomer was identified on the chromatograms with reference to compounds prepared by stereospecific synthesis (tazifylline and ranolazine) or differential salt crystallization (Sotalol)[3]. Optical purity above 95% could be given with a lower limit of 5% of each enantiomer.

EnantioPac columns permit an excellent chiral HPLC resolution of new compounds. For each enantiomeric resolution, several factors should be carefully assessed to optimize this useful routine procedure to monitor optical purity during the asymmetric synthesis.

REFERENCES

1. J. Hermansson, *J.Chromatogr.* 198:67-68 (1984).
2. G. Schill, I. W. Wainer, S. A. Barkan, *J. Liq. Chromatogr.* 9(2&3), 641-666 (1986).
3. A. Simon, J. A. Thomis, Patent No. DE.3.419.067

NOVEL CHIRAL STATIONARY PHASES FOR THE RESOLUTION OF ENANTIOMERS BY LIGAND EXCHANGE CHROMATOGRAPHY (POSTER)

Yoichi Yuki, Kazuhiko Saigo*, Hirokii Kimoto*,
Toru Nishida*, Kouzou Tachibana and Masaki Hasegawa*

Research Center, Daicel Chemical Industries Ltd.
1239 Shinzaike, Aboshi-ku, Himeji, Hyogo 671-12, Japan and
*Department of Synthetic Chemistry, Faculty of Engineering,
The University of Tokyo, Hongo, Bunkyo-ku, Tokyo 113, Japan

The chromatographic resolution of the enantiomers of amino acids and 2-hydroxy acids is conventionally achieved by ligand exchange chromatography using copper (II) complexes of chiral amino acids. Among these CSPs copper (II) complex of L-proline and L-hydroxyproline shows an efficient chiral recognition ability when it is covalently bonded to silica gel pretreated with 3-glycidoxypropyltrimethoxysilane. However, some amino acids such as alanine and glutamic acid could not be resolved on both the CSPs.

Comparing copper (II) complex of L-proline with that of L-hydroxyproline, the chiral recognition ability is somewhat different. This shows that the hydroxyl group in the chiral moiety may play an important role in the chiral recognition. On the other hand, hydro-recognition ability in some CSPs. These facts indicate that the compounds have a hydroxyl group and a hydrophobically interactive site other than chelating site with copper (II) are suitable as CSPs.

CHIRAL STATIONARY PHASES

The design of CSPs was achieved by the use of optically active 2-aminoalkanols, such as D-2-phenylglycinol[1], (-)-norephedrine[2], (1R,2S)-1,2-diphenylethanol[3], (1S,2S)-1,2-diphenylethanol[4], and (R)-2-amino-1,12-triphenylethanol[5]. The CSPs having these amino alcohol moieties were easily prepared as follows: N-Ethoxycarbonlymethylation with ethyl bromcacetate, followed by hydrolysis with Aq. NaOH solution gave mono sodium salts of N-carboxymethylamino-alcohols were then covalently bonded to silica gel pretreated with -glycidoxypropyltrimethoxysilane. The copper (II) complexes were formed by treatment with aqueous copper (II) sulfate solution. The modified silica gel was packed into a stainless column (250mm x 4.6 (id)mm) by slurry method. The resolution of racemates on the copper (II) complexes of these CSPs was carried out by means of ligand exchange chromatography using an aqueous copper (II) sulfate solution as mobile phase.

RESULTS

The CSP's derived from [1] and [5] had little chiral recognition ability for amino acids and their derivatives. The CSP from [3] could resolve a number of the amino acids except for serine threonine, aspartic acid, phenylalanine, and histidine. The CSP from diastereomeric [4] was found to be capable of resolving serine, threonine, phenylalanine, and aspartic acid, although this was not effective for another amino acids. Thus the CSPs derived from [3] and [4], diasteromeric each other, showed significant difference in selectivity for amino acids and also their derivatives, namely, amino acids and their derivatives can be divided into two groups. However, CSP from [2] could resolve amino acids and their deriatives of both groups, but the scope of application and the values obtained are smaller than those showed when [2] and [3] were used as CSPs these differences in the chiral recognition should result from the difference of configuration at C1 and the difference of substituent at C2. In the case of CSPs from [3] and [4], the pi-pi interaction between the phenyl groups at C1 and C2 with hydrobonding of hydroxy groups play an important role to fix the comformation of copper (II) chelates. By contrast, a flexibility of the comformation of CDD from [2] by a lack of the pi-pi interaction would result in the smaller values. Moreover, the cases of CPSs from [1] and [5] indicate that the structure at C1 plays also an important role for the chiral recognition.

MEASUREMENT OF (1) THYROXINE CONCENTRATIONS IN LYOPHILIZED SERUM PREPARATIONS USING PRE-COLUMN DERIVITIZATION WITH O-PHTHALDEHYDE/ N-ACETYL-(1)-CYSTEINE AND CHIRAL SEPARATION BY REVERSE-PHASE HPLC WITH ELECTROCHEMICAL DETECTION (POSTER)

George Lovell and Patrick H. Corran*

Wellcome Research Laboratories, Langley Court
Beckenham, Kent, UK and
*National Institute for Biological Standards and Control
South Mimms, Hertfordshire, UK

Several alternative analytical schemes were evaluated for the analysis of (1) thyroxine concentrations in lyophilized serum preparations intended for use as RIA reference standards.

Reverse-phase HPLC was chosen as the simplest analytical method, and (d) thyroxine appeared to be the best choice as internal standard. A published method which determined (d)/(1) thyroxine ratios directly by EC detection in a chiral system employing CU++, proline, and AG+ eluent was found to have inadequate sensitivity. Suitable thyroxine derivatives were examined to improve selectivity and sensitivity, and a chiral system to separate the Dansyl derivatives of (d) and (1) thyroxine was developed using a Cu++, histidine, and Ag+ chiral eluent with fluorescence detection. However the fluorescence quenching effect of the Cu++ made it difficult to measure physiological concentrations of thyroxine, and problems were encountered with contaminating components eluting in proximity to the thyroxine derivatives. An alternative derivitization with o-phtlaldehyde, using N-acetyl-(1)-cysteine as the participating thiol, was investigated. The derivative was not fluorescent but was found to have electrochemical activity. After derivitization the enantiomeric forms of thyroxine were found to be well separated on a C18 reverse-phase HPLC column using a simple methanol/water eluent. A suitable method based on this derivative was developed to estimate thyroxine concentration in a number of sera and plasma preparations. The specimens were also analyzed by RIA and agreement between the methods was found to be good. The developed method was found to be reproducible (CV%=6.7, n=5) and sensitive (minimum quantifiable amount = 2.5ng) being comparable to the performance of RIA methodologies. However the HPLC approach is inherently more selective in not having the problem of cross-specificity common to thyroxine RIAs.

DETERMINATION OF THE ENANTIOMERS OF α-H AMINO ACIDS, α-ALKYL AMINO ACIDS AND THE CORRESPONDING ACID AMIDES BY MEANS OF HPLC COUPLED WITH AUTOMATED PRECOLUMN DERIVATIZATION (POSTER)

A. Duchateau

DSM Research, Deptartment FA
PO Box 19, 6160 Geleen, The Netherlands

α-Disubstituted amino acids, present in various biologically active systems, have recently received increasing attention, principally because of their activity as enzyme inhibitors. Various methods have been reported for the separation of α-H amino acid enantiomers; however separation of α-disubstituted amino acid enantiomers has not been reported yet. In a few HPLC methods, separation of α-H amino acid enantiomers is established after pre-column derivatization with a chiral reagent. In these methods, no quantitative data are reported. For a good reproducibility, it is essential that the derivatization parameters are automatically controlled. In this report, an HPLC method is described for the quantitative assay of the enantiomers of α-H amino acids, α-disubstituted (α-methyl) amino acids and the corresponding amides.

Amino acids and amides are derivatized with o-phtalaldehyde (OPA) in the presence of a chiral sulfhydryl reagent, N-acetyl-L-cysteine (NAC) [1]. Derivatization and injection are performed automatically.

For chromatography, reversed phase is used as stationary phase; the mobile phase consists of Cu(II), L-Pro and ammonium acetate pH = 6.0. The use of Cu(II) and L-Pro in the mobile phase makes it possible to separate the amino acid enantiomers and the amide enantiomers in the same run under isocratic conditions.

The compounds are detected by a means of a fluorometric detector. A large number of α-H amino acids, α-methyl amino acids and their corresponding amides were tested in the chromatographic system. By simply varying the concentration of organic solvent in the mobile phase, dependent on the amino acid/amide studied, each tested amino acid and amide could be completely separated into their enantiomers. For all tested α-H amino acids and amides, the elution order of the enantiomers was L before D. For all tested α-methyl amino acids and amides, the sequence of elution of the enantiomers was reversed, D before L.

A study has been performed on the reaction kinetics of the reaction of α-methyl amino acid with OPA/NAC. The reaction rate of α-methyl amino acid with OPA/NAC is much lower than the reaction rate of the corresponding α-H amino acid with OPA/NAC. The reaction rate for α-methyl amino acids can be accelerated, by raising the amount of OPA tenfold, while the amount of NAC remains constant. The reproducibility of the method makes it suitable for routine analysis. Coefficient of variation of retention time <0.2%, peak area and peak height <2% (n = 5) for α-methylvaline.

Correlation coefficient from linear regression analysis for α-methylvaline was 0.9999 over a concentration range of more than 1 decade.

REFERENCE

1. N. Nimura, T. Kinoshita, *J. Chromatogr.*, 352:169 (1986).

APPLICATION OF CHIRAL STATIONARY PHASES TO CHIRAL SEPARATIONS USING HPLC (POSTER)

A. Blokland, L. P. C. Delbressine and F. Kaspersen

Organon International BV. Scientific Development Group
PO Box 20, 5340 BH Oss, The Netherlands

HPLC columns with the chiral selector chemically coupled to the stationary phase have been investigated for a number of drug candidates. With these columns - triacetylcellulose/cyclodextrin/α-glycoprotein/Pirkle column/Daicel/Urea — the compounds were chromatographed after optimization of the chromatographic parameters. The number of successful separations was low with R_S values ranging from 0.57-3.30. If a separation was achieved the assay was unsuitable for analytical purposes, e.g. the detection of 0.10% of one enantiomer in mixtures, because of the occurrence of broad chromatographic peaks. The columns are sensitive to small changes in pH and molarity of the eluent and no predictable or consistent chromatographic behavior in a series of related compounds was observed.

However, for the preparative separations of the radiolabelled drugs mianserin and mepirzepine the triacetylcellulose column could be applied successfully. The α-glycoprotein column was used for the study of the pharmacokinetic behavior of both enantiomers of bepridil after administration of the racemic drug.

CHIRAL HPLC STUDIES OF SOME ACIDIC HERBICIDES AND PHARMACEUTICAL ANTI-INFLAMMATORY AGENTS (POSTER)

B. Blessington, N. Crabb*, and J. O'Sullivan*

Department of Pharmaceutical Chemistry
University of Bradford, Bradford BD7 1DP, UK and
*Pharmacy Department, Barnsley General Hospital
Gawber Road, Barnsley, UK

The direct enantiomeric resolution of the racemic herbicide CMPP (2-)4-chloro-2-methylphenoxy)-propanoic acid) was demonstrated on an Enantiopac ($α_1$ acid glycoprotein) chiral HPLC column. The related herbicide, racemic DCPP (2-(2, 4-dichlorophenoxy)-propanoic acid) along with the anti-inflammatory agents ibuprofen and ketoprofen could not be separated on this system (the organic modifier N, N-dimethyloctylamine was not used since its use has been associated with reduced column life).

Various amide derivatives of CMPP, DCPP and ibuprofen were prepared and examined on both the Enantiopac system and a chiral "Ionic Pirkle" comprising of N-(3, 5-dinitrobenzoyl) (R)-(-) phenylglycine as chiral ligand. No separations were observed using the Enantiopac system whilst CMPP, DCPP and ibuprofen were resolved to baseline as diphenylamides on the "Pirkle Ionic" system.

Using optically pure CMPP enantiomers the elution orders were established for the direct resolution on the Enantiopac system and the indirect (as diphenyl amide) resolution on the "Pirkle Ionic" system. The elution order was shown to reverse between the two systems. It was also shown that no racemization occurred during derivatization.

PITFALLS IN THE METHODS FOR ENANTIOMERIC RATIO DETERMINATION OF RACEMIC DRUGS (POSTER)

L. P. C. Delbressine

Organon International BV Scientific Development Group
PO Box 20, 5340 BM Oss, The Netherlands

Increased interest in stereochemical aspects of pharmacological activity and drug disposition has led to a rapidly increasing need of methods for the detection of enantiomers in biological fluids.

Chromatographic separation through a variety of approaches should be the technique of choice but as illustrated for two of our drugs - bepridil and Org 3770 - this is not always without problems. These drugs cannot be converted in a simple way into diastereoisomers while existing HPLC methods for the separation of the enantiomers of these drugs suffer from low sensitivity, selectivity or reproducibility. Alternatives like NMR analysis and the application of pseudoracemates will be discussed.

GC STUDIES OF CHIRAL HERBICIDES AND CHIRAL ANTI-INFLAMMATORY DRUGS (POSTER)

B. Blessington and S. Karkee

Department of Pharmaceutical Chemistry
University of Bradford, Bradford BD7 1DP, UK

A packed-column, achiral, gas-chromatographic method was used to separate and quantify the enantiomers present in the herbicide CMPP (2-(4-chloro-2-methylphenoxy) propanoic acid). The same method was also used to study the enantiomers of the anti-inflammatory drug flurbiprofen (2-(2-fluoro-2-biphenylyl) propanoic acid). In each case the racemic (±) acid was converted to the corresponding amide, using thionylchloride and (+)-1-phenyl-ethylamine. Corresponding reactions with (±)-1-phenyl-ethylamine were also carried out. Authentic samples of each enantiomer of CMPP together with flurbiprofen enantiomers were each individually derivatized under the same conditions. This enabled the elution order of the different diastereoisomeric amides to be defined and also confirmed that negligible racemization occurred during formation of the amides. All products were obtained as crystalline materials and characterized by IR, MS and their GC retention times.

The feasibility of using this method for biological/ecological studies was examined by derivatization and subsequent GC analysis of submicrogram samples. Potential internal standards were investigated: phenoxyacetic acid was selected for CMPP whilst myristic acid was used with flurbiprofen. The linearity and reproducibility of the method was examined.

The "ruggedness" of this method is compared to previously reported direct HPLC methods, employing Enantiopac and "Ionic Pirkle" chiral stationary phases, for the study of the same and related compounds. The significance of this comparison for the routine quality control of commercial processes and end products is discussed.

OPTIMIZATION STRATEGIES IN CHIRAL HIGH-PERFORMANCE LIQUID CHROMATOGRAPHY ON AN α_1-ACID GLYCOPROTEIN COLUMN (POSTER)

T. A. G. Noctor, A. F. Fell and B. Kaye*

Department of Pharmaceutical Chemistry
University of Bradford, Bradford BD7 1DP, UK and
*Pfizer Central Research, Sandwich, Kent, UK

Resolution of enantiomers on the α_1-acid glycoprotein chiral stational phase[1] is highly sensitive to small changes in eluent composition, particularly pH and concentration of organic modifier[1,2]. Single factor optimization procedures can lead to local, rather than global optima, therefore a multifactor strategy is to be preferred. In the present work, the resolution of the enantiomers of oxamniquine [Mansil R.: a Pfizer schistosomicide] was optimized by plotting a chromatographic response function (CRF) against variation in pH and percentage propan-2-ol in aqueous eluent. The CRF algorithm was adapted from that of Berridge[3] as follows:

$$CRF = (R_s / 1.5)^2 - 0.25(T_L - 16) - 0.05(7 - T_F)$$

where R_s = resolution, T_F = time to first peak, T_L = time to last peak.

The formula seeks to provide the best resolution (>1.5) in a total analysis time of less than 16 minutes, where the first peak is well resolved from the solvent front (<7 min). The optimum eluent composition established by this method was:

(10 mM NaH_2PO_4 + 0.1 M NaCl)-Propan-2-ol(99.5:0.5, % w/v); pH* 5.85.

This yielded k' values for the laevorotatory and the dextrorotatory enantiomers of oxamniquine of 1.42 and 2.76, respectively. The detection limit for the leading enantiomer using UV detection at 245 nm was <0.5 ng injected on-column to give SNR = 2. The method is well suited to chiral quality control, but its applicability to drug metabolism studies is yet to be evaluated. A further extension of this approach currently being examined is to explore the 3-dimensional response surface using, for example, the modified simplex[3] to reduce the total number of experiments required.

REFERENCES

1.   Hermansson, J., *J.Chromatogr.*, 298:67-78 (1984).
2.   Noctor, T. A. G., Clarke, B. J., Fell A., F., *Anal.Procs.* 23:441-443 (1986).
3.   Berridge J., *J.Chromatogr.*, 244:1-14 (1982).

NEW CHIRAL STATIONARY PHASES CONTAINING A PHOSPHOROUS ATOM AS ASYMMETRIC CENTER FOR LIQUID CHROMATOGRAPHY (POSTER)

M. Lienne, M. Caude, R. Rosset and A. Tambute*

Laboratoire de Chimie Analytique de l'Ecole Supérieure de Physique et Chimie de Paris, 10 rue Vauquelin, 75231 Paris Cedex 05, France and *Direction des Recherches et Etudes Techniques, Centre d'Etudes du Bouchet, BP no 3, Le Bouchet, 91710 Vert-le-Petit, France

Chiral stationary phases (CSPs) containing asymmetric phorphorous centres were synthetized from enantiomerically pure tertiary phosphine oxides covalently grafted to microparticulate either γ-aminopropylsilanized or γ-mercaptopropylsilanized silica gel. They afforded the resolution of amino acid enantiomers (as their ester or amide derivatives), amines, aminoalcohols and alcohohols as their 3,5-dinitrobenzamides, (3,5-dinitrophenyl) ureas or 3,5-dinitrophenylcarbamates.

For the case of aminoacids (expected for phenylglycine), a base line resolution was achieved on each CSP, with resolution factors varying from 1.5 to 6 and a regular elution order consistent with the absolute configuration of the chiral phosphine oxide grafted on the CSP. Other series of compounds, such as amines and aminoalcohols, did not show a such regular behavior. Chiral recognition mechanisms are proposed which involve the formation of a π–π charge transfer adduct between the π-donor naphthyl group of the phosphine oxide and the π-acceptor 3,5-dinitrophenyl group of the solute, dipolar interaction or hydrogen bonding between complementary basic and acidic sites of the solute and the CSP; additionally repulsive interactions due to steric hindrance around the chiral atoms are advocated. On each CSP 3,5-dinitrobenzamide derivatives were better resolved than their corresponding 3,5-dinitrobenzoylureas. Alcohols carbamates were poorly resolved and alkylamines were not resolved at all.

The chemical structure of the spacer can give some noticeable variations of selectivity and retention time. Spacers containing an amide bond or a hydroxy group can indeed undergo non stereoselective interactions with solutes. Consequently these interactions can induce a decrease of resolution.

Investigations on the anchoring point of the phosphine oxide grafted on the CSP and effects of electron-donating substituents on the naphthyl group of the phosphine oxide provided insight into the chiral recognition mechanisms. Finally we observed that the efficiency and the stability of the CPSs depended greatly on the grafting mode and on the chemical structure of the phosphine oxide. The use of a high percentage of ethanol or chloroform in the mobile phase led for some CSPs to a loss of resolution and a decrease of retention times whereas some other CSPs are not affected and still display symmetrical peaks.

OPTIMIZATION OF THE CHIRAL SEPARATION OF CYTOTOXIC α-METHYLENE LACTONES AND LACTAMES BY LIQUID CHROMATOGRAPHY (POSTER)

M. Lienne, M. Caude, R. Rosset and A. Tambute*

Laboratoire de Chimie Analytique de l'Ecole Supérieure de Physique et Chimie de Paris, 10 rue Vauquelin, 75231 Paris Cedex 05, France and *Direction des Recherches et Etudes Techniques, Centre d'Etudes du Bouchet, BP no 3, Le Bouchet, 91710 Vert-le-Petit, France

The liquid chromatographic resolution of enantiomers of a series of α-methylene γ-lactame and α-methylene lactones was carried out on various commercially available chiral stationary phases (CSPs). A particular interest was given to compounds which exhibit a physiological activity as cytotoxic or antimitotic agents.

On a Pirkle's type CSP prepared by grafting (R)-N-(3,5-dinitrobenzoyl) phenylglycine (DNBPG) covalently to a γ-aminopropyl silica gel, only α-methylene lactames containing two aromatic groups were resolved and the selectivity (α) increased with the π-donor character of the N-aromatic group-hexane/2-propanol and hexane/chloroform mixtures were compared as mobile phases. It is shown that π-π interactions and hydrogen bonding play a role in the chiral recognition mechanism.

Several polymeric CSPs were also studied. (+)Polytriphenylmethylmethacrylate coated on silica gel (Chiralpak 01 (+)) with methanol as eluent afforded rather good values of selectivity (up to 1.8) especially for the case of α-methylene lactones containing aromatic groups; however the efficiency remained low compared to the Pirkle's phase. Effect of the temperature on the resolution was investigated.

The chiral recognition process is probably based on π–π interactions, steric hindrance providing repulsive effects and partial inclusion of the solute in the polymeric layer.

Cellulsoe tribenzoate and trisphenylcarbamate derivatives (resp. Chiracel OB and Chiracel OC) coated on macroporous silica gel displayed selectivities from 1.1 to 1.3; but because of the very poor efficiency of these phases, no baseline resolution was achieved. Solute-CSP diastereomeric interaction may result from the combined effects of π–π interaction, hydrogen bonding and partial inclusion of the solute in cavities of the cellulose derivatives, steric bulkyness at the solute chiral carbon should be therefore considered. Polymeric CSPs seemed to be very sensitive to the geometrical structure of the solutes.

Application of two protein CPs is finally reported: α_1-acid glycoprotein (α_1-AGP) and bovine serum albumin (BSA) immobilized on microparticulate silica gel (resp. Enantiopac and Resolvosil-BSA). The separations were performed using an aqueous phosphate buffer as eluent; retention and selectivity are governed by adding 2-propanol as organic modifier into the mobile phase. Both CSPs exhibited good chiral recognition ability especially towards the cytotoxic solutes with resolution between 1 and 2; here too, due to the very low efficiency of these CSPs, high selectivities are required to achieve a base line separation.

This study demonstrated that it was difficult to find a CSP displaying a high stereo-selectivity and having together a large scale of applications even among a series of compounds. The complexity of the chiral recognition mechanisms involved should be advocated, above all for protein or polymeric CSPs.

DETERMINATION OF THE ENANTIOMERIC COMPOSITION OF SK&F 94836 IN PLASMA AND URINE BY HPLC (POSTER)

N. J. Viney, M. J. Hooper and P. M. Osborne

Physical Organic Chemistry Department
Smith Kline & French Research Ltd
The Frythe, Welwyn, Hertfordshire, UK

SK&F 94836 (2-Cyano-1-methyl-3-[4-(4-methyl-6-oxo-1,4,5,6,-tetrahydropyridazin-3-yl)phenyl[guanidine) has been shown to exhibit positive inotropic and vasodilator activity in animals and is currently under development for the treatment of congestive heart failure. SK&F 94836 contains one centre of optical activity. An assay is described which is capable of the separation and quantitation of the two enantiomers in samples of plasma and urine. Semipreparative HPLC was used to isolate the analyte from interfering species in deproteinated samples.

Separation of the enantiomers was achieved by chiral HPLC on a silica based β-cyclodextrin stationary phase. Validation of the method was performed to ensure that there was no stereoselectivity associated with the assay procedure.

AUTHOR INDEX

Bayer E.	184	Lienne M.	193, 194	
Beesley T.E.	182	Link R.	91	
Béranger S.	186	Lloyd D.K.	131	
Blessington B.	190, 191	Lovell G.	188	
Blokland A.	190	Macaudiere P.	115	
Bridger G.	183	Martin E.	185	
Briggs D.A.	61	Mutschler E.	185	
Burke J.A.III	23	Nishida T.	187	
Caude M.	115, 193, 194	Noctor T.A.G.	121, 192	
Chu Ya-Q.	11	Norinder U.	127	
Clark T.	79	Osborne P.M.	195	
Corran P.H.	188	O'Sullivan J.	190	
Crabb N.	190	Palfreyman	181	
Crooks B.	65	Pirkle W.H.	23	
Dayer P.	71	Pochapsky T.C.	23	
de Zeeuw R.A.	37	Rao N.K.R.	55	
Deas A.H.B.	79	Rosset R.	115, 193, 194	
Delbressine L.P.C.	190, 191	Ruane R.J.	135	
Delée E.	186	Saigo K.	187	
Deming K.C.	23	Schurig V.	91	
Drenth B.F.H.	37	Spahn H.	185	
Duchateau A.	189	Stevenson D.	1	
Feitsma K.G.	37	Stiffin R.M.	11	
Fell A.F.	121, 192	Sundholm E.G.	127	
Gaskell R.M.	65	Tachibana K.	187	
Godfrey R.	61	Tambute A.	193, 194	
Goodall D.M.	131	Tambute T.	115	
Grover P.L.	43	Tandy M.J.	183	
Hall M	43	Taylor D.R.	183	
Hasegawa M.	187	Todd B.	55	
Heintzelmann B.	181	Towill R.C.	55	
Homer R.B.	61	Viney N.J.	195	
Hooper M.J.	195	Vogeler K.	79	
Husband R.	183	Wagner J.	181	
Jullien I.	186	Wainer I.W.	11	
Karkee S.	191	Williams G.	1	
Kaspersen F.	190	Wilson I.D.	135	
Kaye B.	192	Wolf E.	181	
Kimoto H.	187	Yuki Y.	187	
Leeman T.	71	Zreika M.	181	
Le Garrec L.	186			

COMPOUND INDEX

α_1 acid glycoprotein, 14, 38, 122
alanine, PTH derivatives, 137, 138
D-alanine-β-naphthylamide, 138
alaproclate, 127-130
albendazole sulfoxide, 118, 119
allo-isoleucine, 141
alprenolol, 57, 73, 77
amino acid enantiomers, 14, 137
    dansylated amino acids, 61-64, 138, 141
2-aminobutyric acid, 141
α_1 amino glycoprotein, 5, 14
aminoheptane, 117
γ-aminopropyl silanised silica, 45, 47
4-aminophenylalanine, 141
4-aminosalicylic acid, 46
3-amino-3,5,5-trimethyl-butyrolactone-
    HCl, 141
o-anisyl phosphine oxide enantiomers, 117
anthracene, 44, 45-46
anthracene-1,2-dihydrodiol, 43, 47-52
anthracene-1,2-tetrahydrodiol, 47,51, 52
[^{14}C] anthracene, 43
n-aryl-alanine, 23
aspartic acid, 142
atenolol, 65,67
atropine, 39
azole fungicides, 79, 81

BASF 111, 81, 82, 84, 87
benzo(c)phenanthrene (B(c)Ph), 44
benzo(a)pyrene(BP)-7,8-dihydrodiol, 43-54N-
    benzoxycarbonyl-glycyl-L-proline,
    72
(±)2-benzylamino-1 propanol, 6, 65
O-benzylserine, 141
O-benzyltyrosine, 141
betaxolol, 77
R-(+)-betaxolol, 73
bisnaphthol, 39
bitertanol, 81, 82, 83
β-blocking agents, 55-59,71-78
    see also specific names
BP see benzo(a)pyrene
2-bromo-carboxylic acid esters, 105
4-bromophenylalanine, 141

2-bromopropionic acid ethyl ester, 33,
    104-105, 106, 107
5-bromotryptophan, 141
burfuralol, 77
tert-butylmethylcarbinol, 104-105, 107, 111,
    112
tert-butyloxirane, 104-105, 107
tert-butylpropranolol, 56

D-(+)-10-camphorsulfonic acid, 7, 65, 73, 75,
    76
carboxylic acids, 11, 14-15
cellulose esters, 4, 5
chalcogran (E-2-ethyl-1,6-
    dioxaspiro[4,4]nonane), 2, 97,
    99-112
chalcogran (Z-2-ethyl-1,6-
    dioxaspiro[4,4]nonane), 99-112
Z-chalcogran 2, 101
chiral sulfoxides, 17
3-chloralanine, 141
S-(2-chlorobenzyl)-cysteine, 141
m-chloroperbenzoic acid, 93
4-chlorophenylalanine, 141
cobalt(II) camphorate, 91
copper(II)bis[3-(trifluoroacetyl)-(1R)-
    camphorate], 109
cyclodextrins, 4-5, 7, 13, 37
cyclohexylphenylglycolic acid, 39
3-cyclopentylalanine, 141

dansyl (1-dimethylaminonaphthalene-5-
    sulphonyl) amino acids, 61-64,
    138,141
debrisoquine, 75
O-desmethyl-metoprolol
    p-hydroxybenzoate, 72
O-desmethyl-metoprolol, 75, 77
R-(+)-O-desmethyl-metoprolol, 74
dialkylsulfinimines, 17
dialkylsulfoxides, 17
dichlopentezol (S-3308), 79, 81, 87
diclobutrazol, 81, 82, 83, 84
diethoxymethylvinylsilane, 95, 109, 111
trans-1,2-dihydro-1,2-dihydroxyanthracene
    (anthracene-1,2-dihydrodiol), 45

D,L-B-3,4-dihydroxyphenylalanine, 137
(-)-trans-7r,8r-dihydro-7R-
    dihydroxybenzo(a)pyrene (BP-
    7R,8R-dihydrodiol), 43
trans-1,2-dihydroxy-1,2,3,4-tetrahydro-
    anthracene (anthracene-1,2-
    tetrahydriodiol), 47
+(+)-r-7,t-8-dihydroxy-t-9,10-oxy-7,8,9,10
    tetrahydrobenzo(a)pyrene, 43
3,3-dimethylbutan-2-ol (tert-butylmethyl-
    carbinol), 104-105, 107, 111, 112
dimethyl 1,4-dihydro-4-(2-[4-(2-hydroxy-3-
    phenoxy propylamino)]-5-nitro
    phenyl)-2,6-dimethylpyridine-3,5-
    dicarboxylate (M192221), 65
trans-2,3-dimethyloxirane, 106, 107, 112
(±) dinitrobenzamides, alaproclate, 128
3,5-dinitrobenzoylchloride, 37, 128
N-(3,5-dinitrobenzoyl)-α-amino ethyl esters,
    24, 26, 27
N-(3,5-dinitrobenzoyl)-α-amino
    phosphonate methyl esters, 28
N-(3,5-dinitrobenzoyl)-α-amino
    phosphonates, 33
N-(3,5-dinitrobenzoyl)-β-amino esters, 29
N-(3,5-dinitrobenzoyl) alanine amides, 32
N-(3,5-dinitrobenzoyl) leucine amides, 30
N-(3,5-dinitrobenzoyl) phenylglycine
    amides, 31, 37
    diphenyltetramethyldisilazane, 93
disopyramide, 12
1,3-divinyltetramethyldisilazane, 95

enantiomeric amino acids, 11, 121, 137
epibromhydrin, 104, 105, 107
epichlorhydrin, 104-105
(±)epinephrine, 65, 67
'Eraldin' see practolol
N-ethoxycarbonyl-2-ethoxy-1,2-
    dithydroquinoline (EEDQ), 37
E-2-ethyl-1,6-dioxaspiro[4,4]nonane
    (chalcogran E-2), 97, 98-112
Z-2-ethyl-1,6-dioxaspiro[4,4]nonane, 98-112
ethyl phenylglycinates, 33
ethyloxirane, 106, 107

fenarimol, 81
fenpropimorph, 81
(±)-1-ferrocenyl-1-methoxyethane, 138
(±)-1-ferrocenyl-2-methylpropanol, 138
(+)-S-(1-ferrocenylethyl)thioglycolic acid,
    138
2-fluorophenylalanine, 141
3-fluorotyrosine, 141
flutriafol, 81, 82, 84
N-formyl-tert-leucine, 141

glucose, glucose syrup, 132
glutamine, 140, 142
glycoproteins see under A
N-glycylphenylalanine, 141

Z-glycyl-L-proline, 72, 75, 76

heptafluorobutanoylchloride, 94
(3-heptafluorobutanoyl)-(1R)-10-methyl-
    camphor, 94
(3-heptafluorobutanoyl)-(1S)-10-methylene-
    camphor, 94
(3-heptafluorobutanoyl-)(1S)-10-
    dimethoxymethylsilyl)-
    methylcamphor, 94
hexamethyldisilazane, 95
D,L-histidine, 137
homophenylalanine, 141
4-hydroxy-alprenolol, 77
4'-hydroxy-betaxolol, 77
S-(-)-hydroxy-betaxolol, 74
4'-hydroxy-betaxolol fumarate, 72
(2S,4R,2'RS)-4-hydroxy-1-
    (2'hydroxydodecyl)-proline, 140
1'-hydroxy-metoprolol, 77
R-(+)-hydroxy-metoprolol, 74
4'-hydroxy-metoprolol p-hydroxy-benzoate,
    72
4'-hydroxy-metoprolol, 73, 75, 77
S-(-)-hydroxy-metoprolol, 74
cis-4-hydroxyproline, 141
5-hydroxy-propafenone, 77
4-hydroxy-propranolol, 77
hydroxy-1,2,4-triazoles, 79

ibuprofen, 18, 139
'Inderal' see propranolol
4-iodophenylalanine, 141
isoleucine, 138, 140
isopropyloxirane, 104, 106, 107

ketaconazole, 81
(+)-ketamine, 121

lanthanum(III)tris(3-(perfluoroacyl)-(1R)-
    camphorate], 103

M192221, 65,67
maltopentose, 132
maltose, 132
maltotetrose, 132
maltotriose, 132
manganese(II)bis[3-(heptafluorobutanol)-
    (1R)-camphorate], 91, 97
methionine, 138, 141
D,L methionine β naphthylamide, 138
methorphan, 39
4-methoxyphenylalanine, 141
5-methoxytryptophan, 141
± methylatropine, 39
(±)-α-methylbenzylamine, 65, 67
10-methylcamphor, 103
methylcamphorate, 95-109
3-methylcholanthrene (3-MC), 44
2-(1-methylcyclopropyl)-glycine, 141
methyldimethoxysilane, 94

(1S)-10-methylenecamphor, 94, 103
α-methylene γ-lactone, 118, 119
methyloxirane, 104-105, 107, 112
N-methylphenylalanine, 141
2-methyltetrahydrofuran, 104, 105-107, 112
4-methyltryptophan, 141
5-methyltryptophan, 141
6-methyltryptophan, 141
7-methyltryptophan, 141
metoprolol, 71-78
S-(-)-metoprolol tartrate, 72
R-(+)-metoprolol hydrochloride, 72
R-metoprolol tartrate, 72
morpholine fungicides, 79

naphthalene, 44
1-naphthoyl chloride, 128
N-naphthyl α-amino ester, 26
2-naphthyl phosphine oxide, 116, 117
nickel(II)-Chirasil-Metal, 111
nickel(II)bis[(3-heptafluorobutanoyl-)(1S)-
        10-(dimethoxy-methylsilyl)-4-
        nitrophenylalanine, 141
nickel(II)bis[3-heptafluorobutanoyl]-(1R)-
        10-methylene-camphorate], 91-92,
        94,98-112
nickel(II)bis[3-heptafluorobutanoyl]-(1S)-
        10-methylene-camphorate], 94
ninhydrin, 140
norleucine, 141

oxamniquine, 122, 123, 124
oxprenolol, 53, 76, 77
oxyphencyclimine, 39
oxyphenonium bromide, 37-41
oxyprenolol, 53, 57, 76, 77

paclobutrazol [(2RS, 3RS)-1-(4-
        chlorophenyl)-4,4-dimethyl-2-
        (1,2,4-triazol-1-yl)pentan-3-ol], 79,
        80-84, 87
PAH diol-epoxides, 43
PAH metabolites, 43-54
2,3,4,5,6-pentafluorobenzoyl chloride, 128
D,L-phenylalanine, 137, 138, 140
    PTH derivatives, 137
phenylglycine, 141
phenylthiohydantoin amino acid
        enantiomers, 137-143
phosphine oxides, 116
polycyclic aromatic hydrocarbon
        metabolites, 43
    diol epoxides, 43
(+)-polytriphenylmethyl methacrylate CSP,
        119-120
practolol (Eraldine®), 65, 66, 67
proline, 140, 142
promethazine, 39
propafenone, 77
propiconazole, 81

propranolol hydrochloride (Inderal®), 55,
        57, 65, 67, 71, 73, 77
N,N-di-n-propyl-L-alanine, 141
PTH see phenylthiohydantoin amino acid
        enantiomers
pyrimidine fungicides, 79, 81

quinidine, 75

cis, trans roseoxide, 106,107

serine, 137, 138
sodium dodecyl sulphate, 46
sulfimines, 17
sulfinamides, 17
sulfoxides, 11
sulfoximines, 17
sulfur, 14, 17

(+)-tartaric acid, 65
(±)-tartaric acid, 65
'Tenormin', see atenolol
trans-tetrahydro-4-methyl-2-(2-methyl-1-
        propenyl)-pyrane (trans-roseoxide),
        106
tetramethylammonium hydroxide, 95
S-(3-thiabutyl)-cysteine, 141
S-(2-thiapropyl)-cysteine, 141
3-thiazolidine-4-carboxylic acid, 140
D,L-threonine, 137, 138
[14C] triadimefon, 80
triadimefon (1R + 1S), 79, 80, 82, 85, 86, 88
triadimenol, [1-(4-chlorophenoxy)-3,3-
        dimethyl-1-(1,2,4-triazol-1-
        yl)butan-2-ol], 79, 82, 83, 86, 88, 89
triapenthenol, 81, 82, 84, 87
2,2,2-trifluoro-1-(9-anthryl) ethanol, 139
trimethylsilanol, 95
tropic acid, 39
D,L-tryptophan, 137, 138, 140
tyrosine, 138, 140

(R)-urea column, 129

valine, 137, 138, 141

ZGP see N benzoxycarbonyl-glycyl-L-proline

Abstracts, *appendix 3*, 181-195
Achiral separations, 142
Acid-amide derivatization, 16
α_1 Acid glycoprotein column, 15, 122
Acylation, preparation of β-diketonate
    ligands, 94
AGP, *see* α_1 acid glycoprotein
Alaproclate, separation, 127-130
Albumin *see* Bovine serum albumin
Amides, resolution, 139-140
Amino acids
    chiral urea stationary phases, 56
    choice of HPLC-CSP, 14, 16
    dansyl, TLC separation, 135-143
    ligand exchange resolution, 140
    TLC resolution, 137-143
    see also specific compounds
γ-Aminopropyl silanised silica, bound, 45,
    47
Aminoalcohols, resolution by ion-pair
    HPLC, 65-70
Anthracene 1,2-dihydrodiol metabolites
    hydrogenation, 47
    stereoselectivity, 48-49
Anthracene 1,2-tetrahydrodiol, 47, 52
    SS and RR enantiomers, 51
Antiarrhythmic agents, 12
Antidepressant separations, 127
Anti-inflammatory agents, resolution, 18-19
Argon ion-based laser polarimeter, 131
Artificial intelligence and expert systems,
    125

Barley, plant growth regulators, 79-90
Benzamides, resolution, 16
Benzo(a)pyrene 7,8-dihydrodial
    formation by human skin, 47-48
    metabolism by rabbit liver microsomes,
    46
    stereoselectivity, 47-54
    H-3 radiolabelled, metabolism by human
    and rabbit skin, 43, 46-52
Benzyl esters, resolution, 16
β-Blocking agents
    resolution on chiral urea phases, 55-59
    pharmacokinetics, 71

β-Blocking agents (continued)
    underivatised, determination, 71-78
Blood profiles, metoprolol, poor/extensive
    metabolizers, 77
Bovine serum albumin, 5, 14, 15
Botrytis cinerea, 85-87
α-Bromopropionic acid, alkylations, 33
    ethyl ester, 33, 104-105, 107

Camphor *see* 10-methyl camphor
    see also Chiral stationary phases
Capillary zone electrophoresis, 165
Carboxylic acids, CSPs, resolution, 16-17
Carboxylic acid moiety, choosing HPLC-CSP,
    14
Carcinogenicity
    genetic polymorphism, 48
    PAH metabolites, 43, 45
Catalytic hydrogenation, 47, 94
Cellulose, in TLC, 137
Cellulose esters, 5, 6
Cellulose triacetate, microcrystalline, 13
Cellulose tribenzoate, 116-119
Cellulose-based CSPs, 13
Chalcogran, 2
    enantiomer discrimination
        complexation GC, 102
    enantiomer separation, 101
    enantioselectivity, 107
    enthalpy and entropy changes, 102
    retention data, 99
Chiral cavity phases, 4-5
Chiral cyclopropanation catalyst, 109
Chiral discriminating agents, *see* Chiral
    recognition
Chiral gas analysis, fungicides and plant
    growth regulators, 79-90
Chiral ion-pairing HPLC 37-38, 65-70, 71-78,
Chiral ligand exchange chromatography, 13
Chiral mobile phase additives, 7-8
Chiral paramagnetic 'shift' reagents, 1-3
'Chiraplate', 6, 140
Chiral recognition mechanisms, 23-35, 58,
    124
    detectors, suppliers, 15
    historical background, 3

Chiral recognition mechanisms (continued)
    homochiral diastereomeric complex, 33
    circular dichroism, 125
    linear diode-array detection, 125
    polarised light, 125
    three-point rule, 25
Chiral resolution, analytical and
        preparative methods, 65-70
Chiral separations
    artificial intelligence and expert systems,
        125
    development trends, 123-125
    general reviews (appendix), 163
    intermolecular energy calculations, 127-
        130
    iterative lattice, 122
    modified sequential simplex method, 122
    optimisation strategies, 122-123
    optimising, in drug analysis, 121-125
    prediction:, computer aided chemistry,
        127-130
    response surface mapping, 122
Chiral stationary phases
    α_1 acid glycoprotein, 39
    amino acid — ligand exchange, 13
    aqueous liquid membranes, 125
    N-aryl alanine-derived, 23-25
    attractive interactions, 13
    biopolymers, 23-25
    broad-spectrum, 23-25
    Camphorato-Chirametal stationary
        phases, 103
    carboxylic acids, resolution, 16-17
    cellulose-based, 13
    chiral cavity or ravine, 13
    Chiralcel OB, 115-120
    Chiralpak OT(+), 115-120
    Chirametal stationary phases, 91-114
    'Chiraplate', 140
    Chirasil-Metal stationary phase
        complexation GC, 91
        enantioselectivity, 92
        nickel (II), 110-113
        retention-increase and
            enantioselectivity, 104-106
        synthesis, 109-113
        temperature limit, 92
        thermal stability, 92
    Chirasil-Val stationary phases, 6, 92
        structure 7
        triazole separation, 79, 85, 87-88
    classification, 4-7, 13-14
    commercially available, 115-120
    cyclodextrin-based, 13
        see also Cyclodextrins
    derivatization, acid amide, 16
    dipole stacking, 13
    drug analysis, HPLC, 11-21
    for GLC, 6
    for HPLC, 4-6

Chiral stationary phases (continued)
    inclusion complexes, 13
    microcrystalline cellulose triacetate, 13
    phenylmethylacryate polymers, 13
    pi-pi interactions, 13
    Pirkle-type, 115, 118
        see also Pirkle columns
    polysiloxane, 94
    protein based, 14
    semipreparative and upscale separations,
        125
    silica-bound proteins, 5-6
    in supercritical fluid chromatography,
        115-120
    for TLC, 6-7
    types I-V, 13-14
    urea aminoacids, 56
Chiral sulphur compounds, 17-18
Chiral urea phases, 55-59
Chirality
    biological importance, 1-10
    metabolism from non-chiral compounds,
        2
    reagents, 3-4
Chromatographic Response Function, 122
Chromatography see specific procedures
Chromophores, 4
Circular dichroism detection, 125
Cladosporium cucumerinum, 85-87
Cobalt (II) camphorate, 91
Complexation gas chromatography
    enantiomeric separation, 91-114
    thermodynamic aspects, 92, 95-109
    see also Gas liquid chromatography
Computer aided chemistry, predicting chiral
        separations, 127
Conformational analysis 127
Copper, in Chirametal CSPs, 91-114
Copper ions, ligand exchange phase, 5, 6
CSPs see Chiral stationary phases
Cyclobond I, 116-120
Cyclodextrins
    β-acetylated columns, 87
    column CSPs 4-5, 13, 84
    Cyclobond I, 116-117
    -dansyl amino acid complexes, 61-64
    HPLC, 79, 82
    β-cyclodextrins
        TLC applications, 137-138
        TLC, Rf values, 138
Cytochrome P-450
    isozymes, 44, 49
    mono oxygenation, 43
    purified P-450c, 44
Cytochrome P-450b
    anthracene dihydrodiol formation, 49

Dansyl amino acid enantiomers
    complexation by cyclodextrin, 61-64
    fluorescence emission spectra, 61

Derivatising agents
    (-)-methoxyacetyl chloride, 44
    (-)-α-methoxy-α-trifluoromethylphenyl-
        acetyl chloride, 44
Diasteroisomers
    formation, 121
    separation, 136
    transient complexes, 12
Dihydrodiols see Anthracene;,
    Benzo(a)pyrene
β-Diketonate ligands, 94
(L)-N-(3,5-Dinitrobenzoyl)leucine
    aminopropyl TLC, 139
(R)-N-(3,5-Dinitrobenzoyl)phenylglycine
    aminopropyl TLC, 139
DIOL adsorbants, 65
Diphenyltetramethyldisilazane, column, 93
Dipole-dipole interactions, 24
Distomer, definition, 121
Droplet counter-current chromtography,
    appendix, 164
Drug analysis
    HPLC, 11-21
    optimising chiral separations, 121-125
    stereochemical composition, 11-12
Double bond, reduction, 2

Enzyme-mediated biotransformations, 121
Enantiomers
    activity, 121
    definition, 1
    'enantiomeric excess', 91
    iso-enantioselective temperature, 92
    methods of determination, 1-10
    pharmacodynamic parameters, 12
    separation by complexation GC, 91-114
    see also Chiral recognition mechanisms
EnantioPac, 39-41
    see also α1 acid glycoprotein
Epoxidation of oxiranes, 93
Epoxide hydrolase, 43, 44, 49
Eudismic index, 121
Eudismic ratio, 121
Eutomer, definition, 121

Factor analysis, 122
Faraday modulation polarimetry, 131
Fluorescence measurements, dansyl amino
    acid enantiomers, 61-64
Fungicides, 79-90
    foliar application, 85, 88
    seed treatment, 89
Fusarium culmorum, 85-87

Gaeumannomyces graminis, 79, 86-87
Gas-liquid chromatography, 6
    appendix, 158-161
Giberellins, biosynthesis, 79, 87
Glutathione, conjugation, 44

Glycoproteins see α1 acid glycoprotein
    column

Helical polymer phases, 5, 7
High performance liquid chromatography
    β and β-acetylated columns 79-80
    chiral analysis of fungicides and plant
        growth regulators, 79-90
    chiral ion-pairing, 65-70, 71-78
    chiral stationary phases, 4
    complexation of dansyl amino acid
        enantiomers, 61-64
    drug analysis, 11-21
    HPLC-LC, appendix, 147-158
    and mass spectrometry, 20
    optical rotation detector, 131-133
Hydrogen bonding, CSPs, 13
Hydrogenation, catalytic, 47
Hydroxylation, 2

Ibuprofen
    resolution, 18-19
    structure, 19
Immunoassays
    enantiomeric-specific, 8
    steroselective, 71
Intermolecular interaction energies, 128
Iterative lattice, 122

Ketone reduction, 2
Kinetic fractionation, 121
Kupferstecher, principal aggregation
    pheromone, 97

Laser diode, 131
Lattice design, in separations, 122
Ligand exchange chromatography, 5, 13,
    140-143
Linear diode-array detection, 125
Liquid chromotography, chiral separations,
    127-130
Liver microsomal preparations, 43-54, 73

Manganese (II) camphorates, 91
Manufacturers and suppliers, appendix,
    179-180
Methanolysis, CSP polysiloxane, 94
10-Methyl camphor, acylation, 103
Microcrystalline, phenylmethacrylate
    polymers, 13
Microsomal fractions, metabolism of PAH,
    46-54
Modified Sequential Simplex plot, 123-124
Mouse, skin, PAH carcinogenesis, 45, 48-54

(R)-(-)-1-(Naphthyl)ethyl isocyanate, 139-140
Normal phase liquid chromatography, 116
Nuclear magnetic resonance spectrometry, 8
    chiral ion-pair separation, 66
    enantiotopic nuclei, 103

Nuclear magnetic resonance spectrometry
(continued)
nuclear Overhauser effects, 33, 34
proton monitoring, 109

Optical rotation detector for HPLC, 131-133
Organ culture, short term, 46
Oxidation processes, 2
Oxiranes, preparation, 93

Palladium, hydrogenation, 94
Photodiode detector, 131
Pi-pi interactions, 24, 25
Pirkle columns, HPLC, 4, 17, 43-54
    CSPs, 115, 118
    in series with radio HPLC, 85-86
    see also Chiral stationary phases
Pirkle phases, 124
Pityogenes chalcographus, 97
Plant growth regulators, 79-90
Plants, sterol biosynthesis, 79
Platinum catalyst, 94
Polarimeter
    argon ion-based laser, 131
    Faraday modulator, 131
    stopped flow, 131
Polarised light, in chiral detection, 125
Polycyclic aromatic hydrocarbons
    metabolites, 43-54
    sterochemical analyses, 47
(+)-Polytriphenyl methacrylate CSP, 119-
    120
Post-column derivatization, 20
Pre-column derivatization, 18
Preparative chromatography, appendix, 164
Preparative scale separations, 66
Prism polarisers, 131
Protein columns, 5-6, 7

Rabbit, PAH carcinogenesis, 45-46
Racemic drugs, 2
Racemic mixtures, 1
Response surface mapping, 122

Scanning densitometry, 140
Seed treatment, barley, triadimenol
    enantiomers, 85
SFC see Supercritical fluid chromatography
Silica
    γ-aminopropyl silanised, 45, 47
    fused, HR columns, 92, 93
Skin, mammal organ culture, 46-54
Skin culture techniques, 46-54
Stationary phase see Chiral stationary
    phases
Sterol C-14 demethylase, 85
Stopped-flow polarimeter, 131
Subcritical fluid chromatography, 115
Sulphur, chiral compounds, 17-19

Super-critical fluid chromatography, 82,
    115-120
    appendix, 164-165
Suppliers and manufacturers, appendix,
    179-180

(+)-Tartaric acid, 137
Thin layer chromatography
    (-)-alanine, 137
    ascorbic acid, 137
    appendix, 161-163
    C-18 bonded plate, 140-142
    chiral, 135-143
    'Chiral Plate' 6, 140
    β-cyclodextins, 137-138
    multiple developing, 86
    prospects, 135-143
    resolution of metabolites, 46
    in screening before HPLC, 139
    (+)-tartaric acid, 137
    two-dimensional gradient, 142
Triadimenol enantiomers, metabolism after
    seed treatment, 85
Trichlorosilane/H_2PtCl_6, hydrosilylation,
    33

10-Undecenyl alcohol, esterification, 33

L-Valine-tert-butylamide, 82

Xenobiotic compounds, 121